SpringerBriefs in Computer Science

SpringerBriefs present concise summaries of cutting-edge research and practical applications across a wide spectrum of fields. Featuring compact volumes of 50 to 125 pages, the series covers a range of content from professional to academic.

Typical topics might include:

- A timely report of state-of-the art analytical techniques
- A bridge between new research results, as published in journal articles, and a contextual literature review
- A snapshot of a hot or emerging topic
- An in-depth case study or clinical example
- A presentation of core concepts that students must understand in order to make independent contributions

Briefs allow authors to present their ideas and readers to absorb them with minimal time investment. Briefs will be published as part of Springer's eBook collection, with millions of users worldwide. In addition, Briefs will be available for individual print and electronic purchase. Briefs are characterized by fast, global electronic dissemination, standard publishing contracts, easy-to-use manuscript preparation and formatting guidelines, and expedited production schedules. We aim for publication 8–12 weeks after acceptance. Both solicited and unsolicited manuscripts are considered for publication in this series.

**Indexing: This series is indexed in Scopus, Ei-Compendex, and zbMATH **

More information about this series at https://link.springer.com/bookseries/10028

Yixiang Fang • Kai Wang • Xuemin Lin •
Wenjie Zhang

Cohesive Subgraph Search Over Large Heterogeneous Information Networks

Springer

Yixiang Fang
School of Data Science
The Chinese University of Hong Kong,
Shenzhen
Shenzhen, Guangdong, China

Kai Wang
Antai College of Economics & Management
Shanghai Jiao Tong University
Shanghai, China

Xuemin Lin (iD)
Antai College of Economics & Management
Shanghai Jiao Tong University
Shanghai, China

Wenjie Zhang
Computer Science and Engineering
The University of New South Wales
Sydney, NSW, Australia

ISSN 2191-5768   ISSN 2191-5776   (electronic)
SpringerBriefs in Computer Science
ISBN 978-3-030-97567-8   ISBN 978-3-030-97568-5   (eBook)
https://doi.org/10.1007/978-3-030-97568-5

This Springer imprint is published by the registered company Springer Nature Switzerland AG
The registered company address is: Gewerbestrasse 11, 6330 Cham, Switzerland

To the two groups of people who greatly support our research: **our family members** *and* **our collaborators**

Preface

With the advent of a wide spectrum of recent applications, querying large heterogeneous information networks (HINs) has received a great deal of attention from both academic and industrial societies. HINs involve objects (vertices) and links (edges) that are classified into multiple types; examples include bibliography networks, social networks, knowledge networks, and user-item networks in E-business. An important component of these HINs is the cohesive subgraph, or subgraph containing vertices that are densely connected internally. Searching cohesive subgraphs over HINs has been found useful in many real-world applications, such as community search, product recommendation, and fraud detection. Consequently, how to design effective cohesive subgraph models (CSMs) and how to efficiently perform cohesive subgraph search (CSS) on large HINs has become important research topics in the era of big data.

The main purpose of this book is to thoroughly survey the recent technical developments on efficiently performing CSS, in view of the fact that many real graphs are usually large HINs. To have a whole picture of reviewing these works, we classify them according to the classic cohesiveness metrics such as core, truss, clique, connectivity, and density. Meanwhile, since the bipartite network is a special representative type of HINs that can often be processed in a special manner, we also classify HINs into two categories, namely bipartite networks and other general HINs (that are not only customized for bipartite networks). To review these works, we extensively discuss the specific models and their corresponding search solutions.

Moreover, we analyze and compare these CSMs and solutions systematically. Specifically, we first compare different groups of CSMs and analyze their common features and different features from multiple perspectives such as cohesiveness constraints, shared properties, and computational efficiency. Then, for the CSMs in each group, we analyze and compare their model properties (e.g., parameters and constraints) and high-level algorithm ideas. Note that since the bipartite network is a special case of HINs, all the models developed for general HINs can be directly applied to bipartite networks, but the models customized for bipartite networks may not be easily extended for other general HINs due to their restricted settings. Besides, we point out a list of promising research directions in this field.

We believe that this book does not only help researchers to have a better understanding of existing CSMs and solutions but also provides them interesting insights for future research. Consequently, the book can be used either as an extended survey for researchers who are interested in conducting research on cohesive subgraph computation over large HINs, or as a reference book for postgraduate students who are learning courses on the related topics, or as a guideline book for industry engineers to solve real problems using CSS solutions.

Organization The book is organized as follows:

In Chap. 1, we focus on providing the necessary background of CSS over large HINs and giving an introduction to the research field to highlight the popularity and applications in the topic of CSS. More specifically, we first show a list of example HINs (e.g., DBLP, Facebook, and Yago) and typical applications of CSS over HINs (e.g., fraud detection, community search, product recommendation, and biological data analysis). Then, we discuss the challenges of conducting CSS over large HINs. Finally, we make a classification of existing works according to the classic cohesiveness metrics (i.e., core, truss, clique, connectivity, and density).

In Chap. 2, we aim to present the preliminaries of performing cohesive subgraph search. We first formally introduce the data models of HINs and bipartite networks, and then we review typical classic cohesive subgraph models on homogeneous networks, including k-core, k-truss, k-clique, k-edge-connectivity component (k-ECC), and the densest subgraphs.

In Chap. 3, we extensively introduce the five groups of CSMs and solutions for the bipartite networks, which are core-, truss-, clique-, connectivity-, and density-based models and solutions.

In Chap. 4, we extensively introduce the four groups of CSMs and solutions for other general HINs, which are core-, truss-, clique-, and density-based models and solutions. We also review one model that is not covered by the groups above.

In Chap. 5, we perform a thorough analysis and comparison of different CSMs and their corresponding solutions on bipartite networks and other general HINs, respectively, by highlighting their advantages and disadvantages, such as analyzing their computational complexities and application scenarios.

In Chap. 6, we review the two groups of works that are highly related to the topic of our book, which are CSS on homogeneous networks and HIN clustering. In particular, for the first group, we mainly discuss the representative works of five CSMs on conventional homogeneous networks.

In Chap. 7, we discuss several promising future research directions about CSS over HINs, including novel application-driven CSMs, efficient search algorithms, parameter optimization, and an online repository for collecting HIN datasets, tools, and algorithm codes, which can provide researchers with some good starting points to work in this area. In addition, we draw a brief conclusion for the book.

Acknowledgments This book was partially supported by NSFC under grant 62102341, CUHK-SZ grant UDF01002139, National Key R&D Program of China under grant 2018AAA0102502, GuangDong Basic and Applied Basic Research Foundation 2019B1515120048, and Australian Research Council Discovery Projects (DP200101338, DP210101393, DP200101116).

Shenzhen, Guangdong, China   Yixiang Fang
Shanghai, China   Kai Wang
Shanghai, China   Xuemin Lin
Sydney, NSW, Australia   Wenjie Zhang
December 2021

Acknowledgments

This book would not be possible without the guidance of and constant stimulating discussions with my colleagues and fellow researchers at the Chinese University of Hong Kong, Shenzhen, Shanghai Jiao Tong University, and the University of New South Wales. Over the years, our research teams have been jointly funded by multiple government projects sponsored by Australian Research Council Discovery Projects, National Natural Science Foundation of China, National Key R&D Program of China, and GuangDong Basic and Applied Basic Research Foundation. Besides, our research has also been strongly supported by industry companies, such as Huawei Technologies Co., Ltd. and Alibaba Group, and the collaboration with these industry companies on building real applications has led to many of the core findings in this book.

We also would like to sincerely thank our students, whose heavy lifting on many of these research projects has been at least as valuable to us in conducting cohesive subgraph research on large graphs. We also want to thank the collaborators of our research teams for their insightful discussions and great help. Ultimately, without their strong support, this work and its impact would have gone unrealized.

Contents

About the Authors

Yixiang Fang is an associate professor in the School of Data Science, Chinese University of Hong Kong, Shenzhen. He received PhD in computer science from the University of Hong Kong in 2017. After that, he worked as a research associate in the School of Computer Science and Engineering, University of New South Wales, with Prof. Xuemin Lin. His research interests include querying, mining, and analytics of big graph data and big spatial data. He has published extensively in the areas of database and data mining, and most of his papers were published in top-tier conferences (e.g., PVLDB, SIGMOD, ICDE, NeurIPS, and IJCAI) and journals (e.g., TODS, VLDBJ, and TKDE), and one paper was selected as best paper at SIGMOD 2020. He received the 2021 ACM SIGMOD Research Highlight Award. Yixiang is an editorial board member of the journal Information & Processing Management (IPM). He has also served as program committee member for several top conferences (e.g., ICDE, KDD, AAAI, and IJCAI) and invited reviewer for top journals (e.g., TKDE, VLDBJ, and TOC) in the areas of database and data mining.

Kai Wang is an Assistant Professor at Antai College of Economics & Management, Shanghai Jiao Tong University. He received his BSc degree from Zhejiang University in 2016 and his PhD degree from the University of New South Wales in 2020, both in computer science. His research interests lie in big data analytics, especially for the big graph and spatial data. Most of his research works have been published in top-tier database conferences (e.g., SIGMOD, PVLDB, and ICDE) and journals (e.g., VLDBJ and TKDE).

Xuemin Lin is a Chair Professor at Antai College of Economics & Management, Shanghai Jiao Tong University. He is a Fellow of IEEE. He received his BSc degree in applied math from Fudan University in 1984 and his PhD degree in computer science from the University of Queensland in 1992. Currently, he is the editor-in-chief of IEEE Transactions on Knowledge and Data Engineering. His principal research areas are databases and graph visualization.

Wenjie Zhang is a professor and ARC Future Fellow in the School of Computer Science and Engineering at the University of New South Wales in Australia. She received her PhD from the University of New South Wales in 2010. She is an associate editor of IEEE Transactions on Knowledge and Data Engineering. Her research interests lie in large-scale data processing, especially in query processing over spatial and graph/network data.

Acronyms

HIN   Heterogeneous information network
CSS   Cohesive subgraph search
CSM   Cohesive subgraph model
CS   Community search
CD   Community detection
DS   Densest subgraph
DSD   Densest subgraph discovery
MVB   Maximum vertex biclique
MEB   Maximum edge biclique
MBB   Maximum balanced biclique

Chapter 1
Introduction

1.1 Background

With the advent of a wide spectrum of recent applications, querying large hetero-geneous information networks (HINs) has received a great deal of attention from both academic and industrial societies. Unlike conventional homogeneous networks in which vertices are of the same type, HINs [147, 153] involve objects (vertices) and links (edges) that are classified into multiple types. Large HINs are prevalent in various real applications, and here are some well-known typical HINs:

- *DBLP*. As a well-known bibliography network, DBLP includes more than 4.86 million journal articles, conference papers, and other publications on computer science,[1] which describes the relationship among entities of four different types, i.e., paper, author, venue (e.g., conference, journal, etc.), and topic. Figure 1.1a illustrates an HIN of the DBLP network, where the vertices labelled "a", "p", "v", and "t" denote author, paper, venue, and topic, respectively.
- *Facebook*. Facebook is an online social media that has attracted billions of users.[2] The entities (e.g., users and their posted texts, pictures, videos, and events) and their relationship (e.g., friendship and posting relationship) in Facebook naturally form a huge HIN.
- *Yago*. Yago is an open-source knowledge base with over 10 million entities and 120 million facts among these entities,[3] where a fact is a triplet describing the relationship between two entities.
- *User-item network*. In E-commerce platforms (e.g., Alibaba and Amazon), the purchase behaviors between users and products can often be modeled as a

---

[1] https://en.wikipedia.org/wiki/DBLP.

[2] https://en.wikipedia.org/wiki/Facebook.

[3] https://en.wikipedia.org/wiki/YAGO_(database).

© The Author(s), under exclusive license to Springer Nature Switzerland AG 2022

Y. Fang et al., *Cohesive Subgraph Search Over Large Heterogeneous Information Networks*, SpringerBriefs in Computer Science, https://doi.org/10.1007/978-3-030-97568-5_1

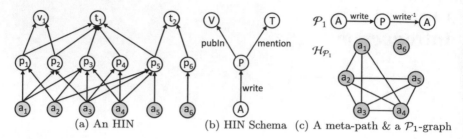

Fig. 1.1 An example HIN of DBLP network [62]

bipartite network, which is a special kind of HIN with only two types of vertices (i.e., users and products).

Cohesive subgraph search (CSS) is a fundamental research topic in network science, which aims to find subgraphs that are densely linked internally from the network. It has been extensively studied on conventional homogeneous networks, and used in various application scenarios [7, 49, 58, 71, 110, 121, 122, 185, 190]. Owing to the prevalence of HINs, how to effectively and efficiently search cohesive subgraphs has become an important research topic in the era of big data. The importance of this topic can be reflected by many real applications [7, 49]. Here are some typical applications of CSS on HINs, to name a few:

- *Fraud detection.* In Alibaba E-commerce platform, the owners of e-business often pay some agents in black markets to promote the rankings of their products. To improve cost effectiveness, these agents often organize a group of users to make fake "purchase" records of a set of products. This will generate many large bicliques in the bipartite graph consisting of users and products. Based on the above observations, researchers of Alibaba Group adopted the maximum biclique model to detect these fraudulent actions and achieved great success [123].
- *Community search.* Recently, the topic of community search (CS) has gained much attention [58] with many real applications (e.g., event organization). As shown in [62, 180], the core- and truss-based cohesive subgraph models (CSMs) are effective for CS over HINs (e.g., finding a community of authors collaborated intensively from the DBLP network).
- *Product recommendation.* In the graph of Instacart co-purchasing which is an HIN, the maximal motif cliques [81] can be used for product recommendation, since the dense subgraph of different grocery items indicates that they are frequently bought together.
- *Biological data analysis.* For example, the relationship between genes and diseases naturally forms a bipartite network. Identifying sets of densely linked genes and diseases from the network can reveal their hidden relationship [51, 136].

1.2 Challenges of CSS Over Large HINs

To search cohesive subgraphs from HINs, there are two key challenges: (1) how to formulate effective CSMs over HINs? and (2) how to develop efficient query solutions?

For the first challenge, researchers often introduce new models by borrowing insights from classic CSMs on conventional homogeneous networks (also called unipartite networks), including k-core, k-truss, k-clique, k-edge-connectivity (also called k-ECC), and the densest subgraph. The formal definitions of these models will be introduced later in Sect. 2. To identify cohesive subgraphs over HINs, a simple idea is to project the HIN into the conventional homogeneous network and then apply the CSMs above to it. This idea, however, may not work since the projected homogeneous network is often much denser than the original HIN; moreover, since vertices and edges of the HIN are multiply typed, carrying different semantic meanings, so it may not make much sense to ignore these types when doing the projection. Therefore, it is better to formulate novel CSMs over HINs directly by considering the key characteristics of HINs, i.e., multiple vertex types and multiple edge types. For example, Fang et al. [62] introduced a novel core model on HIN, which focuses on a specific type of vertices and requires that each vertex is linked to at least k other vertices with the same type via meta-paths (or paths consisting of vertices with multiple types).

The second challenge stems from the fact that real HINs are often with both huge sizes and complicated structures. For instance, as a knowledge base, DBpedia consists of 6 million entities and 9.5 billion RDF triples,[4] and the number of its vertex types[5] is over 685, making it be a huge complicated HIN. To enable efficient CSM, the following several techniques are often exploited in the literature: (1) following the prune-and-verify framework (e.g., [81]) to locate the cohesive subgraphs; (2) building index structures or pre-processing (e.g., [169]) to accelerate the search; (3) designing approximate algorithms that sacrifice some accuracy for achieving higher efficiency; and (4) developing fast parallel algorithms (e.g., [149]).

1.3 Classification of Existing Works of CSS Over HINs

As shown in the literature (e.g., [62, 111, 180, 187, 206]), there are many CSMs and solutions for HINs, but they deal with different types of HINs (e.g., bipartite networks) and formulate CSMs from different angles. Meanwhile, there is a lack of systematic survey of these solutions. Consequently, it is desirable to organize these works and understand how well they perform in terms of efficiency and quality. To

---

[4] https://en.wikipedia.org/wiki/DBpedia.
[5] https://wiki.dbpedia.org/services-resources/ontology.

Table 1.1 Classification of CSMs on HINs

| Models | HINs | |
	Bipartite networks	Other general HINs
Core	(α, β)-core: [5, 23, 48, 111, 112, 170] generalized two-mode core: [23] fractional k-core: [68] τ-strengthened (α, β)-core: [77, 79]	$(k, \mathcal{P})$-cores: [62] r-com: [92] h-structure: [172] $(a_1, \cdots, a_k)$-core: [195] multi-layer core: [65, 66, 113, 203, 204]
Truss	k-bitruss (k-wing): [143, 149, 169, 206] k-tip: [143] quasi-truss: [109]	$(k, \mathcal{P})$-Btruss: [180] $(k, \mathcal{P})$-Ctruss: [180] $(b_1, \cdots, b_k)$-truss: [195]
Clique	maximal biclique: [11, 44, 45, 52, 102, 114, 133, 189] maximum vertex biclique: [47, 78] maximum edge biclique: [47, 123, 137, 146, 151] maximum balanced biclique: [33, 103, 128, 173, 198, 200] quasi-biclique: [1, 89, 118, 130, 150, 177]	maximal motif clique: [81, 101] ABCOutlier-clique: [76] k-partite clique: [47, 74, 115, 129, 139, 195] multi-layer quasi-clique: [19, 20, 138, 183]
Connectivity	k-neighbor connectivity: [97]	–
Density	densest subgraph: [9] (p, q)-biclique densest subgraph: [131]	common densest subgraph: [91] multi-layer densest subgraph: [65, 66]
Others	–	k-context structure: [13]

* Note: $\alpha, \beta, k, p, q, h, a_1, \cdots, a_k, b_1, \cdots$, and b_k are integers; τ is a real value; $\mathcal{P}$ is a meta-path

this end, in this book we will provide a thorough review of these works, in which we will organize, review, and compare different CSMs and solutions, and also point out future research directions in this field.

Specifically, to have a whole picture of reviewing these works, we classify them according to the classic cohesiveness metrics such as core, truss, clique, connectivity, density, etc. Meanwhile, since the bipartite network is a special representative type of HINs that can often be processed in a special manner, we also classify HINs into two categories, namely bipartite networks and other general HINs that are not only customized for bipartite networks. The representative CSMs in each category of networks are listed in Table 1.1, where the ranges of parameters of these models are illustrated in the footnote of the table. To review these works, we will extensively discuss the specific models and their corresponding search solutions. Moreover, we will analyze them systematically and make a fair comparison in terms of both the structure cohesiveness and computational efficiency. Note that since the bipartite network is a special case of HINs, all the models developed for general HINs can be directly applied to bipartite networks, but the models customized for bipartite networks may not be easily extended for other general HINs due to their restricted settings.

We would like to remark that in the literature, there is a highly related group of research works, called HIN clustering or HIN community detection (CD) [147].

Generally, it has similar goals with CSS, but there are three key differences: *(1) The problem settings are different.* CSS aims to search some particular subgraphs satisfying the users' query requests, while HIN clustering focuses on partitioning the entire HIN into a set of dense subgraphs such that each subgraph is weakly linked to others. *(2) The criteria for defining cohesive subgraphs are different.* The criteria for defining cohesive subgraphs are based on query parameters given by the users. In other words, the subgraphs are retrieved depending on user-defined parameters. In contrast, HIN clustering works often use the same global criterion to partition the HIN. *(3) The general approaches are different.* As shown in the literature, CSS solutions can work in an online manner, while HIN clustering works (e.g., [152, 155–157]) are often based on some learning algorithms which are time-consuming and unscalable to large HINs. Due to the major differences above, we will not extensively discuss the works of HIN clustering in this survey.

In summary, our principal contributions are as follows:

- First, we make a classification of existing works according to the classic cohesiveness metrics (i.e., core, truss, clique, connectivity, and density), and then for each class, we review the representative models and solutions thoroughly.
- Second, we perform a thorough analysis and comparison of different CSMs and their corresponding solutions on bipartite networks and other general HINs.
- Third, we offer a list of promising future research directions of CSS over HINs, which can provide researchers some good starting points to work in this area.

Chapter 2
Preliminaries

2.1 Data Models of HINs and Bipartite Networks

Definition 2.1 (HIN [88, 153]) An HIN is a directed graph $\mathcal{H} = (V, E)$ with a vertex type mapping function $\psi : V \to \mathcal{A}$ and an edge type mapping function $\phi : E \to \mathcal{R}$, where each vertex $v \in V$ belongs to a vertex type $\psi(v) \in \mathcal{A}$, each edge $e \in E$ belongs to an edge type (also called relation) $\phi(e) \in \mathcal{R}$, and $|\mathcal{A}| + |\mathcal{R}| > 2$.

Definition 2.2 (Bipartite Network [169]) A bipartite network is an HIN with only two types (layers) of vertices, denoted by $\mathcal{B} = (V = (U, L), E)$, where U (or $U(\mathcal{B})$) denotes the set of vertices in the upper layer, L (or $L(\mathcal{B})$) denotes the set of vertices in the lower layer, $U \cap L = \emptyset$, $V = U \cup L$ denotes the vertex set, and $E \subseteq U \times L$ denotes the edge set.

Clearly, the bipartite network is a special case of the HIN. For HINs that are not only customized for this special case, in this book we call them *other general HINs*. Note that when the network contains only one type of vertices and one type of edges, it is not an HIN; instead, it is a homogeneous network (also called unipartite network), denoted by $G = (V, E)$, where the vertex and edge sets are V and E, respectively.

2.2 CSMs on Homogeneous Networks

Next, we formally present the definitions of classic CSMs on homogeneous networks, including k-core, k-truss, k-clique, k-edge-connectivity component (k-ECC), and the densest subgraphs. We denote the set of neighbors of a vertex v of G by $N_G(v)$. We denote the degree of a vertex v of G by $deg_G(v)$. In the case

© The Author(s), under exclusive license to Springer Nature Switzerland AG 2022
Y. Fang et al., *Cohesive Subgraph Search Over Large Heterogeneous Information Networks*, SpringerBriefs in Computer Science,
https://doi.org/10.1007/978-3-030-97568-5_2

without ambiguity, all the homogeneous networks mentioned later are simple and undirected.

Definition 2.3 (k-**Core** [14, 145]) Given a homogeneous network $G = (V, E)$ and an integer k ($k \geq 0$), the k-core of G, denoted by H_k, is the largest subgraph of G, such that $\forall v \in H_k, deg_{H_k}(v) \geq k$.

Definition 2.4 (k-**Truss** [41, 141, 164, 188]) Given a homogeneous network $G = (V, E)$ and an integer k ($k \geq 2$), the k-truss of G, denoted by J_k, is the largest subgraph of G, such that $\forall e = (u, v) \in J_k, sup(e, J_k) \geq (k-2)$, where $sup(e, J_k)$, or the support of an edge $e=(u, v)\in E$, is $|\{\triangle_{uvw} : u, v, w \in V\}|$, where $\triangle_{uvw}$ denotes a triangle formed by three vertices $u, v, w \in V$.

Definition 2.5 (k-**Clique**) A k-clique is a complete graph with k vertices where there is an edge between every pair of vertices.

Note that k-cores and k-trusses satisfy the *nested property* [14, 58]: given two positive integers i and j, if $i < j$, then $H_j \subseteq H_i$ and $J_j \subseteq J_i$.

Definition 2.6 (Edge Connectivity [70, 82, 83]) Given a homogeneous network $G = (V, E)$ and two vertices $u, v \in V$, the connectivity $\lambda(u, v)$ between u and v is the minimum number of edges whose removal disconnects u and v.

Definition 2.7 (Graph Connectivity [70, 82]) Given a homogeneous network $G = (V, E)$, the connectivity of the graph G, $\lambda(G)= \min_{u,v \in V} \lambda(u, v)$, is the minimum connectivity between any two distinct vertices in G, i.e., the minimum number of edges whose removal disconnects G.

Definition 2.8 (k-**ECC** [70, 82]) Given a homogeneous network $G = (V, E)$ and an integer $k \geq 0$, a subgraph G' of G is a k-edge connected component, or k-ECC, if $\lambda(G') \geq k$ and the connectivity of any subgraph G' satisfying $G \subset G'$ is less than k.

Definition 2.9 (Density [63, 67, 73]) Given a homogeneous network $G = (V, E)$, its density is defined as $\tau(G)= \frac{|E|}{|V|}$.

Definition 2.10 (Densest Subgraph [63, 67, 73]) Given a homogeneous network $G = (V, E)$, its densest subgraph is the subgraph whose density is the highest among all the possible subgraphs.

Example 2.1 Consider the homogeneous network G of Fig. 2.1 and let $G[V]$ denote the subgraph induced by a vertex set V in G. We have: (1) The 0-core and 1-core are G, 2-core is $G[V_1]$ where $V_1 = \{a_2, \cdots, a_6\}$, and 3-core is $G[V_2]$ where $V_2 = \{a_2, \cdots, a_5\}$. (2) The 2-truss is G, 3-truss is $G[V_1]$, and 4-truss is $G[V_2]$. (3) Any triangle is a 3-clique, and $G[V_2]$ is a 4-clique. (4) The 1-ECC is G, 2-ECC is $G[V_1]$, and 3-ECC is $G[V_2]$. (5) The densest subgraph is $G[V_2]$ whose density is $\frac{6}{4}$.

In addition, as pointed out by [58], in terms of structure cohesiveness, the first four metrics above can be roughly ranked as: k-core $\preceq k$-ECC $\preceq k$-truss $\preceq k$-clique, meaning that k-clique is the most cohesive one and k-core is the least cohesive one.

Fig. 2.1 A homogeneous
network G

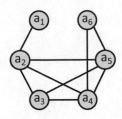

In the meantime, considering the computational efficiency, they can be ranked as:
k-core $\succeq$ k-ECC $\succeq$ k-truss $\succeq$ k-clique, as k-core can be computed in linear time cost
while computing k-cliques is NP-hard.

Chapter 3
CSS on Bipartite Networks

3.1 Core-Based CSMs and Solutions

3.1.1 The (α, β)-Core Model

The notation of (α, β)-core (or (p, q)-core) was first introduced in [5] by extending the classic k-core model to bipartite networks. The definition of (α, β)-core is as follows.

Definition 3.1 ((α, β)-Core) Given a bipartite network $\mathcal{B}$ and degree constraints α and β, a subgraph $R_{\alpha,\beta}$ is the (α, β)-core of $\mathcal{B}$ if (1) $deg_{R_{\alpha,\beta}}(u) \geq \alpha$ for each vertex $u \subset U(R_{\alpha,\beta})$ and $deg_{R_{\alpha,\beta}}(v) \geq \beta$ for each vertex $v \in L(R_{\alpha,\beta})$; (2) $R_{\alpha,\beta}$ is maximal, i.e., any supergraph $\mathcal{B}' \supset R_{\alpha,\beta}$ is not an (α, β)-core.

For instance, considering the bipartite network $\mathcal{B}$ in Fig. 3.1, the $(1, 3)$-core is the subgraph induced by the set of vertices $\{u_1, u_2, u_3, u_4, u_5, u_6, v_3, v_4, v_5\}$. In [5], Ahmed et al. presented a case study to analyse the Internet Movie Database (IMDB) with visualization, and the (α, β)-core model was used to identify meaningful and important subgraphs. Ding et al. [48] integrated the (α, β)-core model in the subspace-clustering-based group recommendation system to improve the efficiency. In their work, an online search algorithm was proposed to retrieve the vertex set of the (α, β)-core by given a bipartite network $\mathcal{B}$, and parameters α and β. By iteratively removing the vertices with degrees less than α in $U(\mathcal{B})$ and the vertices with degrees less than β in $L(\mathcal{B})$, the algorithm can compute the (α, β)-core in $O(m)$ time.

Note that the online search algorithm proposed in [48] needs to traverse the whole bipartite network once (in the worst case). Considering the bipartite network can be very large and the queries to compute the (α, β)-core can be requested frequently in real-world scenarios, this solution is inefficient. Alternatively, one may consider precomputing all the (α, β)-cores and indexing the vertices of them through

© The Author(s), under exclusive license to Springer Nature Switzerland AG 2022
Y. Fang et al., *Cohesive Subgraph Search Over Large Heterogeneous Information Networks*, SpringerBriefs in Computer Science,
https://doi.org/10.1007/978-3-030-97568-5_3

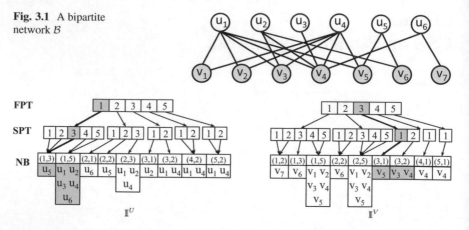

Fig. 3.1 A bipartite network $\mathcal{B}$

Fig. 3.2 The BiCore-Index [111]

a two-level pointer tables (recording α and β). Then, the vertex set of the (α, β)-core can be obtained in the optimal time cost by retrieving the vertices in the location referred by α and β. However, this approach needs $O(n^3)$ space to index all the (α, β)-cores and it is also impractical for handling large bipartite networks. Motivated by the above issues, Liu et al. [111] proposed the BiCore-Index to support the optimal retrieval of the vertex set of the (α, β)-core which needs only $O(m)$ space.

The BiCore-Index $\mathbb{I}$ consists of two parts (i.e., $\mathbb{I}^U$ and $\mathbb{I}^V$) and each of them is constructed from two levels of pointer tables (FPT and SPT) and node blocks (NB). The BiCore-Index of the bipartite network $\mathcal{B}$ in Fig. 3.1 is shown in Fig. 3.2. For example, if the $(1, 3)$-core is requested, we follow the pointer kept in the first element in $\mathbb{I}^U$.FTP and the third element of the 1st array in $\mathbb{I}^U$.STP to obtain u_5 in the node block $(1, 3)$. We continue to visit the node block $(1, 5)$ and stop at node block $(2, 1)$ since 2 is larger than 1. Thus, we obtain $\{u_1, u_2, u_3, u_4, u_5, u_6\}$ from $\mathbb{I}^U$. Similarly, we obtain $\{v_3, v_4, v_5\}$ from $\mathbb{I}^V$. The principal intuition behind the BiCore-Index is that the (α, β)-cores follow a hierarchical structure (i.e., (α, β)-core $\subseteq (\alpha', \beta')$-core if $\alpha \geq \alpha'$ and $\beta \geq \beta'$) and we do not need to store all the (α, β)-cores separately. In addition, utilizing this hierarchical property, the BiCore-Index can be constructed in $O(\delta \cdot m)$ by computing all the (α, β)-cores with $\alpha \leq \delta$ and all the (α, β)-cores with $\beta \leq \delta$. Here δ is the maximal value where the (δ, δ)-core exists in $\mathcal{B}$, and δ is bounded by $\sqrt{m}$. In [112], Liu et al. also studied the incremental maintenance algorithms to maintain the BiCore-Index efficiently when the network is dynamically changing.

In recent works [170, 171], Wang et al. proposed an index to support optimal retrieval of the connected component of the (α, β)-core containing a query vertex q. In addition, they studied the community search problem on weighted bipartite graphs where the edges are associated with weight values. In their model, the (α, β)-core is used to ensure the structure cohesiveness of the community. The community search problem on vertex-weighted bipartite networks was studied in [191] based on the (α, β)-core model. In addition, the (α, β)-core minimization problem was investigated in [29].

3.1.2 The Generalized Two-Mode Core Model

Based on the k-core model in unipartite networks, the generalized core model [15] was proposed which uses a vertex property function instead of vertex degrees to consider different properties of vertices. By combining the ideas from the generalized core in unipartite networks and the (α, β)-core in bipartite networks, the generalized two-mode core [23] is defined as follows.

Definition 3.2 (Generalized Two-Mode Core) Consider a bipartite network $\mathcal{B} = (V = (U, L), E)$, two vertex property functions f and g, and degree constraints α and β. Let $P(V)$ be a power set of the set V. Let f and g be two vertex property functions defined on the network $\mathcal{B}$: $f, g: V \times P(V) \to \mathbb{R}_0^+$. A subset of vertices $C \subseteq V$ in $\mathcal{B}$ is a generalized two-mode core if and only if in the subgraph induced by C, it holds that (1) $f(u, C) \geq \alpha$ for each $u \in C \cap U(G)$, and $g(v, C) \geq \beta$ for each $v \in C \cap L(G)$; (2) C is maximal, i.e., any supergraph $G' \supset C$ is not a generalized two-mode core.

According to the above definition, we list some example vertex property functions below. (1) $f_1(v, C) = deg_C(v)$ which is the degree of a vertex v within C. When this function is used, the generalized two-mode core is exactly the (α, β)-core. (2) For weighted bipartite networks where each edge (u, v) has a weight $w(u, v)$, we can choose $f_2(v, C) = \sum_{u \in N_C(v)} w(v, u)$ which is the sum of weights of edges within C that have a vertex v as an end vertex.

The algorithms to obtain the generalized two-mode core for monotonic and local vertex property functions were also proposed in [23]. The idea is to repeatedly remove the vertices that do not belong to C, which is similar to the online search algorithm for retrieving the (α, β)-core. In addition, a binary heap implementation of priority queues is used to organize the vertices. In this manner, it can efficiently get the vertex with the smallest value of the property function as the root element in a heap.

Note that, a function $f(v, C)$ is said to be *local* iff

$$f(v, C) = f(v, N_C(v)) \text{ for all } v \in V(G) \text{ and } C \subseteq V(G). \tag{3.1}$$

The function is *monotonic* iff

$$C_1 \subset C_2 \Rightarrow \forall v \in V(G) : f(v, C_1) \leq f(v, C_2). \tag{3.2}$$

3.1.3   The Fractional k-Core Model

In [68], Giatsidis et al. proposed the fractional k-core model to evaluate communities in the DBLP bipartite network. The main idea is to project the bipartite network $\mathcal{B} = (V = (U, L), E)$ into a weighted unipartite network $\mathcal{H}$ using a weighting function. The weighted projection network $\mathcal{H}$ is defined as follows.

Definition 3.3 (Weighted Projection) Given a bipartite network $\mathcal{B} = (V = (U, L), E)$, the weighted projection network $\mathcal{H}$ of $\mathcal{B}$ is defined as:

$$\mathcal{H} = (U(\mathcal{B}), \{\{u, u'\} | \exists v \in L(\mathcal{B}) : \{u, v\}, \{u', v\} \in E(\mathcal{B})) \tag{3.3}$$

In addition, for each edge $e = (u, u')$ on $\mathcal{H}$, we set its edge weight to

$$w(e) = \sum_{v \in N_\mathcal{B}(u) \cap N_\mathcal{B}(u')} \frac{1}{N_\mathcal{B}(v)} \tag{3.4}$$

Based on Definition 3.3, the vertex fractional degree of a vertex in $\mathcal{H}$ can be defined as follows.

Definition 3.4 (Vertex Fractional Degree) Given a weighted projection network $\mathcal{H}$ and a vertex $u \in \mathcal{H}$, its fractional degree is computed as

$$fdeg_\mathcal{H}(u) = \sum_{u' \in N_\mathcal{H}(u)} w((u, u')) \tag{3.5}$$

Definition 3.5 (Fractional k-Core) Given a weighted projection network $\mathcal{H}$ and a real value $k \geq 0$, the fractional k-core of $\mathcal{H}$ is the maximal subgraph of $\mathcal{H}$ where each vertex has a fractional degree of k or more.

Essentially, the work [68] is about fractional cores on weighted unipartite networks where the edges and the edge weights are generated from the bipartite networks. This kind of approach is also called projection in the literature, and it may lead to many drawbacks such as edge explosion and information loss [136, 143].

3.1.4   The τ-Strengthened (α, β)-Core Model

In [77, 79], the authors proposed a CSM called τ-strengthened (α, β)-core (or (α, β, ω)-core) which considers both tie strength and vertex engagement on bipartite networks. The definition of τ-strengthened (α, β)-core is based on the following concepts.

Definition 3.6 (Butterfly) In a bipartite network $\mathcal{B}$, given vertices $u, w \in U(\mathcal{B})$ and $v, x \in L(\mathcal{B})$, a butterfly $\bowtie$ is the complete subgraph induced by $\{u, v, w, x\}$, which means both u and w are connected to v and x by edges. The total number of butterflies in $\mathcal{B}$ is denoted as $\bowtie_{\mathcal{B}}$. The total numbers of butterflies containing an edge e and a vertex v are denoted by $\bowtie_e$ and $\bowtie_v$, respectively.

Note that (α, β)-core is a vertex-induced subgraph model, which assumes that the edges are of equal importance. However, in real-world scenarios, the edges can have different tie strength. On bipartite networks, the tie strength (or butterfly support) of an edge e can be modeled as the number of butterflies containing e (i.e., $\bowtie_e$).

Definition 3.7 (Strong Tie) Given a strength level $\tau \geq 0$, an edge $e \in E(\mathcal{B})$ is called a strong tie if $\bowtie_e \geq \tau$. In addition, an edge e with $\bowtie_e < \tau$ is called a weak tie.

Definition 3.8 (Engagement) Given a strength level τ and $u \in V(\mathcal{B})$, the engagement $eng(u)$ is the number of strong ties incident to u. Note that at strength level 0, $eng(u) = deg_{\mathcal{B}}(u)$.

Definition 3.9 (τ-Strengthened (α, β)-Core) Given a bipartite network $\mathcal{B}$, engagement constraints α and β, and strength level τ, a subgraph C is the τ-strengthened (α, β)-core of $\mathcal{B}$, if (1) $eng(u) \geq \alpha$ for each $u \in U(C)$ and $eng(v) \geq \beta$ for each $v \subset L(C)$; and (2) C is maximal.

The algorithms to find the τ-strengthened (α, β)-core were studied in [77, 79]. In [77], a peeling-based online algorithm was proposed, which first computes the (α, β)-core of $\mathcal{B}$ since the τ-strengthened (α, β)-core is always a subgraph of the (α, β)-core, and then computes the supports of edges and the engagements of vertices in the (α, β)-core, and iteratively deletes the vertices without enough engagements. Obviously, this approach is not inefficient to handle large-scale bipartite networks, so the index-based approaches were proposed in [79]. Three partial indexes $I_{\alpha, \beta}$, $I_{\beta, \tau}$, and $I_{\alpha, \tau}$ that selectively store the τ-strengthened (α, β)-cores for some α, β, and τ combinations are proposed. In addition, a feed-forward neural network is trained to predict the best choice of the partial index that minimizes the query time.

3.2   Truss-Based CSMs and Solutions

3.2.1   The k-Bitruss Model

The k-bitruss model is formulated based on the butterfly structure by extending the classic k-truss model on the unipartite network. Notice that butterfly is the smallest non-trivial biclique in bipartite networks which is recognized as an analogue of triangle in unipartite networks [168]. In [206], Zou et al. first introduced the k-bitruss model as follows.

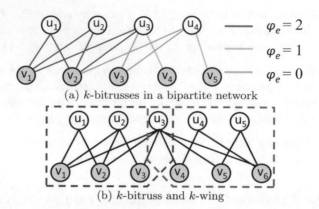

(a) k-bitrusses in a bipartite network

(b) k-bitruss and k-wing

Fig. 3.3 Illustrating the k-bitruss and k-wing

Definition 3.10 (k-Bitruss) Given a bipartite network $\mathcal{B}$ and a positive integer k, the k-bitruss of $\mathcal{B}$ denoted as H_k is the maximal subgraph of $\mathcal{B}$ where $\mathbb{X}_e \geq k$ for each edge $e \in H_k$.

Definition 3.11 (Bitruss Number) Given a bipartite network $\mathcal{B}$, the bitruss number of an edge e denoted as ϕ_e is the largest k such that there is a k-bitruss in $\mathcal{B}$ containing it.

By applying the butterfly connectedness condition to distinguish different regions that are traditionally connected, the authors in [143] proposed the k-wing model. A k-wing is a subgraph of k-bitruss which further needs each edge pair (u_1, v_1) and (u_2, v_2) in the k-wing is connected by a series of butterflies.

Example 3.1 Figure 3.3a shows the bitruss number of each edge. Note that the bitruss number of the edges in blue color is 2, and these edges form the 2-bitruss. In addition, Fig. 3.3b illustrates the difference between k-bitruss and k-wing. The whole network is a 2-bitruss since each edge is contained in at least 2 butterflies. Note that the edges in the left part cannot connect to the edges in the right part through butterflies. For example, (u_3, v_3) and (u_3, v_4) are not connected via butterflies. Thus, there exist two separate 2-wings in red and blue color as shown in Fig. 3.3b)

Given a bipartite network $\mathcal{B}$, to obtain all the k-bitrusses (k-wings) with $k \geq 0$, we need to compute the bitruss number for each edge $e \in \mathcal{B}$ which is also known as the *bitruss decomposition* problem.

In [143, 206], the authors proposed bottom-up approaches to solve the bitruss decomposition problem, which have the worst-case time complexity of $O(m^2)$. The algorithms first compute the butterfly support for each edge and then iteratively peel the edges with the lowest butterfly supports. Note that when an edge e is removed in the peeling process, we need to update the butterfly supports of the (affected) edges which share butterflies with e correspondingly. To achieve this goal,

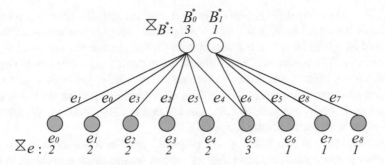

Fig. 3.4 An example of the BE-index

when removing an edge e, the algorithms in [143, 206] need to enumerate all the butterflies containing e.

When handling large-scale bipartite networks, a huge number of edges need to be removed and enumerating all the butterflies containing each removed edge is inefficient. In [169], Wang et al. proposed an online index (i.e., the BE-index) to accelerate the process of updating the butterfly supports of the affected edges when removing an edge. The intuition of the proposed index is to compact the butterflies using a more cohesive structure (i.e., the *bloom* structure which is a biclique with exactly 2 vertices in one layer). In addition, to make sure that each butterfly is contained in exactly one bloom, we only need to find the maximal priority-obeyed blooms (denoted as B^*) which is the maximal bloom where the largest-priority-vertex in it belongs to the layer with exactly two vertices. Note that the higher the degree, the higher the priority; and the ties are broken by the IDs of vertices. Utilizing the BE-index, when we remove an edge e, all the affected edges can be directly found through the blooms without enumerating all the butterflies containing e. Below we give an example of the BE-index.

Example 3.2 For the bipartite network in Fig. 3.3a, the BE-index I can be constructed as shown in Fig. 3.4. We denote the edges (u_1, v_1), (u_1, v_2), (u_2, v_1), (u_2, v_2), (u_3, v_1), (u_3, v_2), (u_3, v_3), (u_4, v_2), (u_4, v_3) by e_0, e_1, e_2, e_3, e_4, e_5, e_6, e_7, e_8, respectively. B_0^* is the subgraph induced by the vertex set $\{u_1, u_2, u_3, v_1, v_2\}$ and B_1^* is the subgraph induced by $\{u_3, u_4, v_2, v_3\}$. In $U(I)$, $\mathcal{X}_{B^*}$ is recorded which is the number of butterflies contained in B^*. In $L(I)$, $\mathcal{X}_e$ is recorded (e.g., $\mathcal{X}_{e_0} = 2$). In $E(I)$, the twin edges are recorded (e.g., the twin edge of e_0 in B_0^* is e_1). Note that for an edge e in B^*, its twin edge will be contained in no butterfly in B^* after removing e. Suppose the edge e_6 is removed, we can find 3 affected edges (i.e., e_5, e_7 and e_8) through B_1^* in I. Since e_5 is the twin edge of e_6 in B_1^*, we update $\mathcal{X}_{e_5}$ to 2. Since the butterfly supports of the edges e_7 and e_8 are equal to $\mathcal{X}_{e_6} = 1$, their butterfly supports do not need to be updated.

Based on the BE-index, a bottom-up peeling algorithm was proposed in [169], which iteratively removes the edges with the minimal support. The algorithm has a

$O(\sum_{(u,v)\in E(\mathcal{B})} min\{deg(u), deg(v)\} + \bowtie_\mathcal{B})$ time complexity. Here $\bowtie_\mathcal{B}$ is the total number of butterflies in $\mathcal{B}$. In addition, a progressive compression approach was also proposed in [169] based on the observation that $\bowtie_e$ is a lower bound of ϕ_e for a given edge e. The algorithm has a different edge processing order comparing with the bottom-up algorithms. It first processes a bunch of edges with high butterfly supports and then compresses these edges after processing them. In this way, the number of butterfly support updates can be significantly reduced, especially for these edges with high butterfly supports.

Furthermore, Shi and Shun [149] proposed parallel algorithms based on the algorithm proposed in [143] to solve the bitruss decomposition problem in multi-thread environments. The main idea is to remove the edges with the same (minimal) butterfly support in one iteration in a parallel way. In addition, a more efficient parallel solution was proposed in [167] recently based on the sequential algorithm in [169]. The personalized k-wing search problem was studied in [2] to find k-wings containing a specific query vertex.

3.2.2 The k-Tip Model

Based on the butterfly structure, the k-tip model is proposed as follows.

Definition 3.12 (k-Tip [143]) Given a bipartite network $\mathcal{B}$ and a positive integer k, a subgraph of $\mathcal{B}$ is a k-tip of $\mathcal{B}$ denoted as T_k, if (1) $\bowtie_v \geq k$ for each vertex $v \in U(T_k)$; (2) each vertex pair $(u, v) \in U(T_k)$ is connected by series of butterflies; and (3) T_k is maximal.

The difference between k-tip and k-bitruss is that k-tip measures the intensity of vertex engagement in the butterfly structure and k-bitruss (or k-wing) focuses on the edge. For example, the subgraph in red color is the 2-tip of the given bipartite network as shown in Fig. 3.5 since each upper vertex in it is contained in at least 2 butterflies.

Definition 3.13 (Tip Number) Given a bipartite network $\mathcal{B}$, the tip number of a vertex v is the largest k such that there is a k-tip in $\mathcal{B}$ containing it.

Similar to the bitruss decomposition, the *tip decomposition* aims to find the tip numbers of all the upper-layer vertices in a bipartite network. In [143], the authors proposed a peeling-based tip decomposition algorithm to compute the tip

Fig. 3.5 Illustrating k-tip

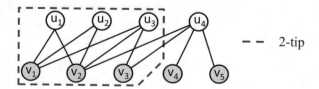

Fig. 3.6 Illustrating the transformation (the special edges are marked in red color)

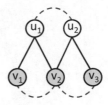

numbers in $O(n^2)$ time. Recently, the authors in [98] propose a coarse-to-fine parallel framework to solve the k-tip decomposition problem efficiently.

3.2.3 The Quasi-Truss Model

In [109], Li et al. proposed the quasi-truss model. The intuition of this model is generating special edges (i.e., triangles) for bipartite networks. For two vertices (in the same layer) of the bipartite network, a special edge is added between them if they have at least one common neighbor. After such transformation, the network has triangle structures as shown in Fig. 3.6. Then, we can discover the quasi-truss based on the number of triangles containing a special edge. For instance, the maximum quasi-truss is the subgraph in which a special edge e' is contained in the maximal number of triangles.

3.3 Clique-Based CSMs and Solutions

3.3.1 The Maximal Biclique Model

The maximal biclique model is a well-known CSM on biparitite networks which is defined as follows.

Definition 3.14 (Maximal Biclique) Given a bipartite network $\mathcal{B}$, a subgraph C is a maximal biclique if (1) for each vertex pair $u, v \in U(C) \times L(C)$, $(u, v) \in E(C)$; and (2) C is maximal.

Example 3.3 Consider the bipartite network $\mathcal{B}$ in Fig. 3.7. The subgraphs induced by $\{u_1, u_2, u_3, v_1, v_2, v_3\}$ and $\{u_5, v_3, v_4, v_5, v_6, v_7, v_8\}$ are two maximal bicliques.

Based on the maximal biclique model, the maximal biclique enumeration problem which aims to find all the maximal bicliques in a bipartite network was studied in many recent works [3, 11, 44, 45, 102, 114, 133, 189]. As shown in [52], the maximal biclique enumeration problem cannot be solved in polynomial time.

Das and Tirthapura [45] proposed the state-of-the-art parallel algorithms to solve this problem. The main framework of their algorithms is similar to the sequential

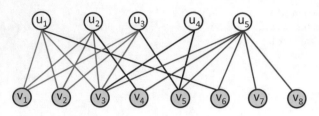

Fig. 3.7 Illustrating the biclique models

algorithm proposed in [114] (i.e., MineLMBC). The MineLMBC algorithm enumerates all the maximal bicliques of a bipartite network B by traversing B in a
depth-first manner. The key improvement of their proposed solutions is that before
enumerating all bicliques containing a vertex, the search space is reduced to the 2-
hop neighbors of the vertex. In addition, to enable effective parallelization, the load
distribution is considered by assigning different rank values to the vertices according
to their degrees.

In [44, 127], the authors also proposed algorithms to incrementally maintain
the maximal bicliques for dynamic bipartite networks where the edges are added
over time. In [11], Ban and Duan studied the maximal half isolated biclique model
that is considered as a variant of the maximal biclique model, and aimed to find
a subgraph in which the edges are incident to outside vertices from at most one
layer. The k-biplex model was introduced in [182], where each vertex of one layer
is disconnected from at most k vertices of the other layer. In [179], the authors
studied the (p, q)-biclique enumeration problem.

There are also three well-known variants based on the maximal biclique model
which are the maximum vertex biclique (MVB), the maximum edge biclique
(MEB) and the maximum balanced biclique (MBB). We introduce these models
and corresponding solutions in the following parts.

3.3.2   The Maximum Vertex Biclique Model

Based on the maximal biclique model, the maximum vertex biclique is defined as
follows.

Definition 3.15 (Maximum Vertex Biclique) Given a bipartite network B, a
subgraph C is the maximum vertex biclique (MVB) if (1) it is a maximal biclique;
and (2) there is no other maximal biclique C' with $|V(C')| > |V(C)|$.

Example 3.4 Consider the bipartite network B in Fig. 3.7. The subgraph induced by
$\{u_5, v_3, v_4, v_5, v_6, v_7, v_8\}$ is the maximum vertex biclique in B since it contains the
maximum number of vertices.

Given a bipartite network B, the MVB in B can be found in polynomial time [78]. As shown in [47], finding the MVB is formulated as an instance of the integer linear programming problem. Note that finding the MVB in B is equivalent to finding a minimum vertex cover in the bipartite complementary network of B. In [78], the authors proposed that finding a minimum vertex cover in a bipartite network can be reduced to finding a maximum matching in the bipartite network which can be solved in $O(nm)$ time by using the well-known Ford-Fulkerson algorithm. Thus, we can find a MVB in a given bipartite network in polynomial time.

3.3.3 The Maximum Edge Biclique Model

The maximum edge biclique is defined as follows.

Definition 3.16 (Maximum Edge Biclqiue) Given a bipartite network B, a subgraph C is the maximum edge biclique (MEB) if (1) it is a maximal biclique; and (2) there is no other maximal biclique C' with $|E(C')| > |E(C)|$.

Example 3.5 Consider the bipartite network B in Fig. 3.7. The subgraph induced by $\{u_1, u_2, u_3, v_1, v_2, v_3\}$ is the MEB in B since it contains the maximum number of edges.

In the literature, given a bipartite network B, many works focus on searching the MEB in B (i.e., the MEB problem). In [137], the authors proved that the MEB problem is NP-hard. The integer-programming-based algorithms were discussed in [47, 151]. Shaham et al. [146] proposed an approximate algorithm to find the MEB based on subspace clustering. Specifically, the algorithm extracts a set of maximal bicliques which contains a maximum biclique under a fixed probability.

The state-of-the-art exact algorithm to find the MEB was proposed in [123] which can handle large bipartite networks. The algorithm follows a progressive manner to find the optimal MEB C^*. Firstly, the algorithm initializes a maximal biclique C_0^* and the parameter τ_L^0. Then, multiple τ_U^k and τ_L^k threshold pairs are used to guess the lower bound of $|U(C^*)|$ and $|L(C^*)|$, and the values of τ_U^k and τ_L^k are progressively updated to increase their product. For each pair τ_U^k and τ_L^k, it first obtains a small subgraph with reduction techniques and applies the branch-and-bound algorithm in [189] to compute a maximal biclique C_k^* with larger size. After running multiple iterations, the algorithm is guaranteed to find the MEB.

Besides, many works studied the MEB problem in some special cases of bipartite graphs including convex bipartite graphs [134] and tree convex bipartite graphs [30]. In addition, the variants of the MEB problem were studied in [4, 135, 159]. In [135, 159], the authors studied the maximum weighted edge biclique problem on weighted bipartite graphs. Specifically, the maximum weighted edge biclique problem is to find a biclique such that the sum of its edge weights is maximized. The problem is proved to be NP-hard and is inapproximable in [159]. Interestingly, Pandey et al. [135] found that it is $O(n^2)$-time solvable for bipartite permutation graphs and is $O(m + n)$-time solvable for chain graphs. In [4], the MEB packing problem was studied.

3.3.4   The Maximum Balanced Biclique Model

The maximum balanced biclique is defined as follows.

Definition 3.17 (Maximum Balanced Biclique) Given a bipartite network $\mathcal{B}$, a subgraph C is the maximum balanced biclique (MBB) if (1) it is a balanced biclique, i.e., $|U(C)| = |L(C)|$; and (2) there is no other balanced biclique C' with $|U(C')| + |L(C')| > |U(C)| + |L(C)|$.

Example 3.6 Consider the bipartite network $\mathcal{B}$ in Fig. 3.7. The subgraph induced by $\{u_1, u_2, u_3, v_1, v_2, v_3\}$ is the MBB in $\mathcal{B}$.

Given a bipartite network $\mathcal{B}$, heuristic approaches [8, 72, 103, 158, 173, 198] were proposed to find the MBB in $\mathcal{B}$. In [8, 158], the authors showed that the MBB search problem is equivalent to the maximum independent set problem, and the algorithms for solving the maximum independent set problem are used to find the MBB. [198] used a tube-search-based method [72] to solve the MBB search problem and designed dedicated graph reduction strategies. A local search framework with several heuristics was studied in [173] and a swap based algorithm was proposed in [103].

Zhou et al. [200], McCreesh and Prosser [128], Chen et al. [33] studied the algorithms to find the exact MBB. Branch and bound algorithms were proposed in [128, 200] with designed pruning rules. In [33], the authors proposed a novel $O^*(1.3803^n)$ algorithm, called denseMBB, by exploiting the fact that finding the MBB is solvable within polynomial time when the input bipartite network satisfies specific constraints.

3.3.5   The Quasi-Biclique Model

As the biclique model is too strict in many real-world applications, the quasi-biclique models were investigated in the literature. A quasi-biclique is considered as a relaxation of the biclique model which allows to miss a certain number of edges (and vertices) from a biclique. By considering different kinds of error-tolerances, there exist many quasi-biclique models in the literature including γ-quasi-biclique [1, 89], ϵ-quasi-biclique [130], symmetrical and balanced ϵ-quasi-biclique [150], δ-quasi-biclique [118], and α-quasi-biclique [177].

To characterize these kinds of error-tolerances, two notations *symmetrical* and *balanced* can be used [150]. We say an error-tolerance is symmetrical if vertices in both layers can tolerate missing edges in the quasi-biclique. We say an error-tolerance is balanced if vertices in both layers can tolerate missing up to the same threshold of edges in the quasi-biclique. By defining an error-tolerance that is symmetrical and balanced, we can ensure each vertex in a layer is closely related to the vertices in the opposite layer in a quasi-biclique.

By considering the subgraph density, the γ-quasi-biclique is defined as follows.

Definition 3.18 (γ-Quasi-Biclique) Given a bipartite network $\mathcal{B}$ and $\gamma \in (0, 1]$, a subgraph C is a γ-quasi-biclique if $|E(C)| \geq \gamma \cdot |U(C)| \cdot |L(C)|$.

Clearly, the error-tolerance in the γ-quasi-biclique is not balanced but symmetrical. In [1], the authors aim to find the γ-quasi-biclique with large cardinality by using a greedy function. In [89], Ignatov et al. proposed mixed integer programming based algorithms to search the maximal γ-quasi-biclique.

Mishara et al. [130] provided a different definition of the quasi-biclique as follows.

Definition 3.19 (ϵ-Quasi-Biclique) Given a bipartite network $\mathcal{B}$ and $\epsilon \in (0, 1]$, a subgraph C is a ϵ-quasi-biclique if each vertex in $U(C)$ is adjacent to at least $(1-\epsilon)$ of the vertices in $L(C)$.

Note that the above ϵ-quasi-biclique is neither symmetrical nor balanced. This is because in a ϵ-quasi-biclique, there is no error-tolerant requirement on vertices in the lower layer. Motivated by this, Sim et al. [150] proposed the symmetrical and balanced ϵ-quasi-biclique model as follows.

Definition 3.20 (Symmetrical and Balanced ϵ-Quasi-Biclique) Given a bipartite network $\mathcal{B}$ and an integer ϵ, a subgraph C is a symmetrical and balanced ϵ-quasi-biclique if each vertex in $U(C)$ is adjacent to at least $|L(C)| - \epsilon$ vertices in $L(C)$, and each vertex in $L(C)$ is adjacent to at least $|U(C)| - \epsilon$ vertices in $U(C)$.

Another similar model as the symmetrical and balanced ϵ-quasi-biclique was proposed by Liu et al. [118] which is called the δ-quasi-biclique.

Definition 3.21 (δ-Quasi-Biclique) Given a bipartite network $\mathcal{B}$ and $\delta \in (0, 0.5]$, a subgraph C is a δ-quasi-biclique if each vertex in $U(C)$ is adjacent to at least $(1-\delta)$ of the vertices in $L(C)$, and each vertex in $L(C)$ is adjacent to at least $(1-\delta)$ of the vertices in $U(C)$.

Note that the δ-quasi-biclique applies inherently the same error-tolerance as the symmetrical and balanced ϵ-quasi-biclique. In addition, Yan et al. [177] introduced the α-quasi-biclique which is defined as follows.

Definition 3.22 (α-Quasi-Biclique) Given a bipartite network $\mathcal{B}$, a maximal biclique $C' \subseteq \mathcal{B}$, and $\alpha \in (0, 1]$, a subgraph C is a α-quasi-biclique if (1) $C' \subseteq C$; (2) each vertex in $U(C)\backslash U(C')$ is adjacent to at least α of the vertices in $L(C')$; and (3) each vertex in $L(C)\backslash L(C')$ is adjacent to at least α of the vertices in $U(C')$.

Note that the error-tolerance of the α-quasi-biclique is symmetrical but not balanced since a α-quasi-biclique is expanded from a maximal biclique.

3.4   Connectivity-Based CSMs and Solutions

Considering a bipartite network with users on the upper layer and interests on the lower layer, users with k common interests are more connected than users with $k-1$ common interests. Motivated by this, Kumar et al. [97] proposed the k-neighbor connectivity plot (or KNC-plot) which can reflect bipartite networks' macroscopic connectivity.

Definition 3.23 (KNC-Plot) Given a bipartite network $\mathcal{B}$ and an integer $k \geq 1$, the two vertices in the pair $(u, u') \in U$ in $\mathcal{B}$ are said to be k-neighbors in $\mathcal{B}$ if there are distinct vertices $v_1, \cdots, v_k \in L$ such that $(u, v_i) \in E$ and $(u', v_i) \in E$ for every $i = 1, \cdots, k$. The k-neighborhood graph $G_k = (U, E_k)$ is a graph such that $(u, u') \in E_k$ if and only if (u, u') are k-neighbors in $\mathcal{B}$. The KNC-plot shows statistics of the component structure of G_k as a function of k.

Example 3.7 Consider the bipartite network $\mathcal{B}$ in Fig. 3.8. The k-neighborhood graphs G_1, G_2 and G_3 for $k = 1, 2, 3$ are also shown in Fig. 3.8.

To compute G_k, the authors of [97] first proposed two naive solutions named Alg-Intersect and Alg-Tuple. The main idea of Alg-Intersect is to check if the size of the intersection of neighbor sets is at least k for any two vertices in $U(\mathcal{B})$. The algorithm Alg-Tuple constructs the k-tuples of neighbors (i.e., combinations of k neighbors) for each $u \in U(\mathcal{B})$. Then, by sorting or hashing the tuples, it can determine whether two vertices in $U(\mathcal{B})$ share a same tuple. Note that, Alg-Tuple is efficient when all the degrees of vertices in $(\mathcal{B})$ are small. Furthermore, a hybrid algorithm, called Alg-Hybrid, which combines the two naive solutions was proposed. Alg-Hybrid partitions the vertex set $U(\mathcal{B})$ into two sets according to the degrees of vertices. Specifically, it calls Alg-Intersect to handle the vertices with high degrees while runs Alg-Tuple to handle the rest of vertices with small degrees.

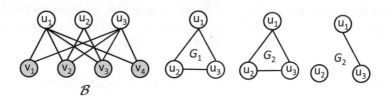

Fig. 3.8 Illustrating the KNC-plot

3.5   Density-Based CSMs and Solutions

3.5.1   The Densest Subgraph Model

In [94], Kannan and Vinay proposed a density measure for bipartite networks which is defined as follows.

Definition 3.24 (Bipartite Graph Density)   Given a bipartite network $\mathcal{B}$, its bipartite graph density is defined as $d(\mathcal{B}) = \frac{|E|}{\sqrt{|U||L|}}$.

Note that the only difference between Definition 2.9 (i.e., the general graph density function) and Definition 3.24 is the denominator. In addition, the function in Definition 3.24 is considered as a weaker objective function than the function in Definition 2.9 [94].

Based on the density definition in Definition 3.24, the bipartite densest subgraph of a bipartite network is defined as follows.

Definition 3.25 (Bipartite Densest Subgraph)   Given a bipartite network $\mathcal{B}$, the bipartite densest subgraph of $\mathcal{B}$ is the subgraph which has the highest bipartite graph density among all the possible subgraphs in $\mathcal{B}$.

Example 3.8   Consider the bipartite network in Fig. 3.9. The bipartite densest subgraph in this network is the subgraph induced by $\{u_1, u_2, u_4, v_1, v_2, v_3, v_4, v_5\}$, and the density of the bipartite densest subgraph is $\frac{10}{\sqrt{3 \times 5}} \approx 2.58$.

In [9], Andersen proposed an algorithm for finding the densest subgraph containing a specific query vertex with a target size of k. The algorithm is considered a local exploration algorithm since it examines only a small subset of the whole network. Andersen also proved that the running time of this algorithm is $O(\Delta k^2)$, where Δ is the maximum vertex degree. Note that the density measurement in Definition 3.24 was originally defined in the setting of directed graphs [94] rather than bipartite graphs. In [9], Andersen theoretically proved that this density measurement is equivalent to the density of directed graphs. Thus, the algorithms for discovering the densest subgraph on directed graphs (e.g., [124]) can be applied to compute the bipartite densest subgraph.

Fig. 3.9   An example bipartite network

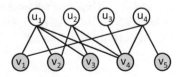

3.5.2   The (p, q)-Biclique Densest Subgraph Model

In [131], the authors proposed the (p, q)-biclique density model to measure the density of a subgraph in the bipartite network. A (p, q)-biclique is a biclique with exactly p upper vertices and q lower vertices.

Definition 3.26 $((p, q)$-**Biclique Density**) Given a bipartite network $\mathcal{B}$ and two integers $p, q \geq 1$, its (p, q)-biclique density $d_{p,q}(\mathcal{B})$ is defined as $d_{p,q}(\mathcal{B}) = \frac{c_{p,q}(\mathcal{B})}{|V|}$, where $c_{p,q}(\mathcal{B})$ is the number of (p, q)-bicliques in $\mathcal{B}$.

According to the above definition, the (p, q)-biclique densest subgraph of a bipartite network can be defined as follows.

Definition 3.27 $((p, q)$-**Biclique Densest Subgraph**) Given a bipartite network $\mathcal{B}$ and two integers $p, q \geq 1$, the (p, q)-biclique densest subgraph of $\mathcal{B}$ is the subgraph which has the highest (p, q)-biclique density among all the possible subgraphs in $\mathcal{B}$.

Example 3.9 Consider the bipartite network in Fig. 3.9. If p and q are set to 2, then the $(2, 2)$-biclique densest subgraph in this network is the subgraph induced by the vertex set $\{u_1, u_2, v_2, v_3, v_4\}$ which has a $(2, 2)$-biclique density of $\frac{3}{5} = 0.6$.

To obtain the (p, q)-biclique densest subgraph, the authors transformed the optimization problem into a decision problem (i.e., to decide whether a (p, q)-biclique densest subgraph with density $\geq D$ exists). Then, the (p, q)-biclique densest subgraph can be obtained using the binary search method. To solve the decision problem, a flow network F is constructed from the original bipartite network. Then, the decision problem is transformed to compute the minimum st-cut in F by applying Gusfield's algorithm [6]. Besides, the randomized approximate algorithms were proposed in [131] to handle large-scale bipartite networks.

3.6   Conclusions

In this chapter, we have focused on the bipartite network, and reviewed the core-, truss-, clique-, connectivity-, and density-based models and the solutions to search subgraphs following these models. Finding these subgraphs can benefit many key applications in bipartite networks such as user-item [36], people-location [100], and author-paper [165]. An in-depth comparison analysis of these models and solutions will be performed in Chap. 5.

Chapter 4
CSS on Other General HINs

4.1 Some Key Concepts on HINs

Definition 4.1 (HIN Schema [88, 153]) Given an HIN $\mathcal{H} = (V, E)$ with mappings $\psi : V \to \mathcal{A}$ and $\phi : E \to \mathcal{R}$, its schema $\mathcal{T}_{\mathcal{H}}$ is a directed graph defined over vertex types $\mathcal{A}$ and edge types (as relations) $\mathcal{R}$, i.e., $\mathcal{T}_{\mathcal{H}} = (\mathcal{A}, \mathcal{R})$. Note that if there is an edge R from vertex type A to vertex type B, the inverse edge R^{-1} naturally exists from B to A.

Definition 4.2 (Meta-Path [153]) A meta-path $\mathcal{P}$ is a path defined on an HIN schema $\mathcal{T}_{\mathcal{H}} = (\mathcal{A}, \mathcal{R})$, and is denoted in the form $A_1 \xrightarrow{R_1} A_2 \xrightarrow{R_2} \cdots \xrightarrow{R_l} A_{l+1}$, where l is the length of $\mathcal{P}$, $A_i \in \mathcal{A}$, and $R_i \in \mathcal{R}(1 \leq i \leq l)$.

Essentially, the HIN schema describes all allowable edge types between vertex types in the HIN, where each edge type can describe one-to-one, one-to-many, or many-to-many relationship. In cases that there exist no multiple edges between the same pair of vertex types, a meta-path can also be simply written in a sequence of vertex types, e.g., $\mathcal{P} = (A_1 A_2 \cdots A_{l+1})$. A path $p = a_1 \to a_2 \cdots \to a_{l+1}$ between vertices a_1 and a_{l+1} is called a *path instance* of $\mathcal{P}$, if $\forall i$, the vertex a_i and edge $e_i = (a_i, a_{i+1})$ satisfy $\psi(a_i) = A_i$ and $\phi(e_i) = R_i$. Note that we use lower-case letters (e.g., a_1) to denote vertices in an HIN, and upper-case letters (e.g., A) to denote vertex types. We say that a vertex u is a $\mathcal{P}$-*neighbor* of a vertex v, or (u, v) is a $\mathcal{P}$-pair, if they can be connected by an instance of $\mathcal{P}$. We say that two vertices u and v are $\mathcal{P}$-*connected*, if there exists a chain of vertices from u to v, such that any vertex is a $\mathcal{P}$-neighbor of its adjacent vertex in the chain.

Definition 4.3 ($\mathcal{P}$-Graph [58, 180]) Given an HIN $\mathcal{H} = (V, E)$ and a symmetric meta-path $\mathcal{P}$, the $\mathcal{P}$-graph is a homogeneous graph $\mathcal{H}_{\mathcal{P}}$, which contains all the vertices with the end type of $\mathcal{P}$ and each pair of vertices is linked by an edge if they are form a $\mathcal{P}$-pair in $\mathcal{H}$.

© The Author(s), under exclusive license to Springer Nature Switzerland AG 2022
Y. Fang et al., *Cohesive Subgraph Search Over Large Heterogeneous Information Networks*, SpringerBriefs in Computer Science,
https://doi.org/10.1007/978-3-030-97568-5_4

For example, Fig. 1.1b shows the schema of DBLP network, where the vertices labelled "A", "P", "V", and "T" denote author, paper, venue, and topic, respectively. In Fig. 1.1c, the meta-path $\mathcal{P}_1$, defined on authors (A) and papers (P), describes two authors with co-authorship, which can also be written as $\mathcal{P}_1 = (APA)$. Since its reverse meta-path is still $\mathcal{P}_1$, it is a symmetric meta-path. In Fig. 1.1a, the path $a_1 \rightarrow p_1 \rightarrow a_2$ is a path instance of $\mathcal{P}_1$, which implies that a_1 is a $\mathcal{P}_1$-neighbor of a_2, and they form a $\mathcal{P}_1$-pair. The $\mathcal{P}_1$-graph $\mathcal{H}_{\mathcal{P}_1}$ is also depicted in Fig. 1.1c.

Two important special types of general HINs are the k-partite network [74, 115, 139] and multi-layer network [64, 203, 204]. A k-partite network ($k \geq 2$) has k disjoint sets of vertices and each edge only connects two vertices from two different sets, and it is often used to model the relationship among k disjoint sets of objects. A multi-layer network (a.k.a. multi-dimensional network or multi-view network) has only one type of vertices, but multiple types of edges with each edge type denoting a specific layer (e.g., a set of users is often involved in multiple social networks simultaneously, so their friendships among these social networks can be modeled as a multi-layer network).

Definition 4.4 (k-Partite Network) A k-partite network is represented by $\mathcal{H} = (V = (V_1, V_2, \cdots, V_k), E)$ satisfying (1) $V = \cup_{i=1}^{k}$; (2) $\forall i, j \in [1, k]$ with $i \neq j$, $V_i \cap V_j = \emptyset$; and (3) $\forall e = (u, v) \in E$, vertices u and v are from two different vertex sets V_i and V_j respectively. Each vertex set is also called a partite set. Note that when $k = 2$, the k-partite network is the bipartite network.

Definition 4.5 (Multi-Layer Network) A multi-layer network is denoted by $\mathcal{H} = (V, E = (E_1, E_2, \cdots, E_l))$, where V is the set of vertices with the same type, E_i ($i \in [1, l]$) is the set of edges with the i-th edge type, and l is the number of layers or edge types. Note that the i-th layer forms a homogeneous (unipartite) network, denoted by $G_i = (V, E_i)$.

4.2   Core-Based CSMs and Solutions

In the literature, there are several HIN core models, namely $(k, \mathcal{P})$-core [62], r-com [92], h-structure [172], $(a_1, \cdots, a_k)$-core [195], and multi-layer core [65, 66, 113, 203, 204].

4.2.1   The (k, $\mathcal{P}$)-Core Models

In [62], Fang et al. proposed three kinds of $(k, \mathcal{P})$-cores, namely basic, edge-, and vertex-disjoint $(k, \mathcal{P})$-cores, which use a symmetric meta-path $\mathcal{P}$ to characterize the cohesiveness of a set of vertices with the same type. The type linked by $\mathcal{P}$ is called target type.

Given a vertex v and a set S of vertices with target type in an HIN, the number of $\mathcal{P}$-neighbors of v within S is called b-degree, denoted by $\alpha(v, S)$. Based on the concept of b-degree, the basic $(k, \mathcal{P})$-core model is defined as follows.

Definition 4.6 (Basic $(k, \mathcal{P})$-Core) Given an HIN $\mathcal{H}$, an integer k, and a symmetric meta-path $\mathcal{P}$, a basic $(k, \mathcal{P})$-core of $\mathcal{H}$ is a maximal set $\mathbf{B}_{k,\mathcal{P}}$ of $\mathcal{P}$-connected vertices, s.t. $\forall v \in \mathbf{B}_{k,\mathcal{P}}$, $\alpha(v, \mathbf{B}_{k,\mathcal{P}}) \geq k$, where vertices of $\mathbf{B}_{k,\mathcal{P}}$ are with the type linked by $\mathcal{P}$.

For example, in Fig. 1.1a, let $\mathcal{P} = (APA)$. Then, the vertex set $C_1 = \{a_1, a_2, \cdots, a_5\}$ forms a $(3, \mathcal{P})$-core. Although this model is straightforward, it may lead to some vertices weakly engaged in the core. Consider the author a_5 in the example above. Although it has three co-authors, a_5 publishes only one paper p_5, while each other author has three papers. In other words, a_5 may be a junior researcher, but is included in a core of senior researchers. This is because the edge "$a_5 \rightarrow p_5$" is shared by three path instances of $\mathcal{P}$. This issue is even more serious for long meta-paths [62].

To tackle this issue, the vertex- and edge-disjoint $(k, \mathcal{P})$-core models are developed. Specifically, consider a vertex v with the target type and let $\Psi[v]$ be a set of path instances of $\mathcal{P}$ starting from v. If $\forall p_1, p_2 \in \Psi[v]$, their i-th $(1 \leq i \leq l)$ edges are different and $(l + 1)$-th vertices are different, then $\Psi[v]$ is a set of *edge-disjoint* paths. Similarly, if $\forall p_1, p_2 \in \Psi[v]$, their i-th $(2 \leq i \leq l + 1)$ vertices are different, then $\Psi[v]$ is a set of *vertex-disjoint* paths. Note that when the length of path equals to 2, a vertex-disjoint path is also an edge-disjoint path. For example, in Fig. 1.1a, let $\mathcal{P} = (APA)$. Then, by using edge-disjoint (or vertex-disjoint) paths, a_1 can be connected to three vertices, while a_5 is connected to at most one vertex, as depicted in Fig. 4.1 (each path is represented by a specific kind of dashed line). Let $\beta(v, S)$ (resp., $\gamma(v, S)$), also called e-degree (resp., v-degree), denote the maximum number of edge-disjoint (resp., vertex-disjoint) path instances starting from a vertex v and ending at vertices in a vertex set S with the target type. Then, the two new core models are formulated:

Definition 4.7 (Edge-Disjoint $(k, \mathcal{P})$-Core) Given an HIN $\mathcal{H}$, an integer k, and a symmetric meta-path $\mathcal{P}$, an edge-disjoint $(k, \mathcal{P})$-core of $\mathcal{H}$ is a maximal set $\mathbf{E}_{k,\mathcal{P}}$ of $\mathcal{P}$-connected vertices s.t. $\forall v \in \mathbf{E}_{k,\mathcal{P}}$, $\beta(v, \mathbf{E}_{k,\mathcal{P}}) \geq k$, where vertices of $\mathbf{E}_{k,\mathcal{P}}$ are with the type linked by $\mathcal{P}$.

Definition 4.8 (Vertex-Disjoint $(k, \mathcal{P})$-Core) Given an HIN $\mathcal{H}$, an integer k, and a symmetric meta-path $\mathcal{P}$, a vertex-disjoint $(k, \mathcal{P})$-core of $\mathcal{H}$ is a maximal set $\mathbf{V}_{k,\mathcal{P}}$ of

Fig. 4.1 Edge- and
vertex-disjoint paths [62]

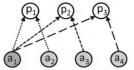

(a) Disjoint paths from a_1 (b) Disjoint path from a_5

$\mathcal{P}$-connected vertices s.t. $\forall v \in \mathbf{V}_{k,\mathcal{P}}$, $\gamma(v, \mathbf{V}_{k,\mathcal{P}}) \geq k$, where vertices of $\mathbf{V}_{k,\mathcal{P}}$ are with the type linked by $\mathcal{P}$.

For example, in Fig. 1.1a, let $v = a_6$, $S = \{a_1, \cdots, a_6\}$, and $\mathcal{P} = (APA)$. Then, $\alpha(v, S) = \beta(v, S) = \gamma(v, S) = 1$. Besides, for $(k, \mathcal{P})$-cores, we have $\mathbf{B}_{1,\mathcal{P}} = \mathbf{B}_{2,\mathcal{P}} = \mathbf{B}_{3,\mathcal{P}} = \{a_1, \cdots, a_5\}$, $\mathbf{E}_{1,\mathcal{P}} = \mathbf{V}_{1,\mathcal{P}} = \{a_1, \cdots, a_5\}$, and $\mathbf{E}_{2,\mathcal{P}} = \mathbf{V}_{2,\mathcal{P}} = \mathbf{E}_{3,\mathcal{P}} = \mathbf{V}_{3,\mathcal{P}} = \{a_1, a_2, a_3, a_4\}$. Note that there is no $\mathbf{B}_{4,\mathcal{P}}$, $\mathbf{E}_{4,\mathcal{P}}$, or $\mathbf{V}_{4,\mathcal{P}}$.

Given k and $\mathcal{P}$, to compute the basic $(k, \mathcal{P})$-core, a simple approach is to first build a $\mathcal{P}$-graph and then compute the k-core of $\mathcal{P}$-graph [14] as the result. However, the $\mathcal{P}$-graph is often much denser than $\mathcal{H}$, especially when $\mathcal{P}$ is very long. As a result, another faster approach is to extract the core from the HIN directly, by iteratively peeling vertices whose b-degrees are less than k [62].

The computation of edge-disjoint $(k, \mathcal{P})$-core is more challenging, as it cannot be computed from $\mathcal{P}$-graphs; instead, it can only be computed from the HIN. A key step of computing it is to compute the e-degree $\beta(v, S)$ of a vertex v. As shown in [62], an exact algorithm of computing $\beta(v, S)$ needs to first build a flow network, and then use the max-flow algorithm to derive the maximum number of edge-disjoint paths. An approximation algorithm with theoretical guarantee is also presented, which iteratively finds an instance of $\mathcal{P}$ and removes its edges from the HIN. Based on the algorithms of computing $\beta(v, S)$, the edge-disjoint $(k, \mathcal{P})$-core can be computed by iteratively peeling vertices with $\beta(v, S) < k$ from the HIN. The algorithm of computing the vertex-disjoint $(k, \mathcal{P})$-core is similar to the above algorithm, except that we need to slightly adapt the algorithms of computing $\beta(v, S)$ for computing $\gamma(v, S)$. In addition, for a specific $\mathcal{P}$, all the $(k, \mathcal{P})$-cores of an HIN can be compactly organized into a tree index where the details can be found in [62].

4.2.2   The r-Com Model

In [92], Jian et al. proposed a novel HIN core model, called r-com. Unlike the $(k, \mathcal{P})$-core which focuses on a set of vertices with the same type, the r-com is defined as a subgraph of vertices with multiple types.

To establish the cohesiveness relationship between two vertex types, a concept of *relational constraint* is introduced. A relationship constraint, denoted by a triplet $\langle l_1, l_2, k \rangle$, describes that each vertex with type l_1 must have at least k neighbors with type l_2, where $l_1, l_2 \in \mathcal{A}$. In Fig. 4.2, for example, two constraints on the DBLP network are shown, where the first one means "each author published ≥ 2 papers", and the second one says "each paper has ≥ 3 authors". Since a single constraint

Fig. 4.2 Two relational constraints [92]

models the relationship between two vertex types, a set of such constraints can be used to model the relationship among multiple vertex types. Based on this idea, the r-com model is developed.

Definition 4.9 (r-Com) Given an HIN $\mathcal{H}$, a set S of relational constraints, a connected subgraph $R = (V_R, E_R)$ is an r-com defined by S, if and only if $\forall v \in V_R$,

- $\psi(v) \in L_S$, where $\psi(v)$ is the vertex type of v, and L_S denotes the set of vertex types included by the constraints in S;
- v is qualified, i.e., $\forall \langle l_1, l_2, k \rangle \in S$, one of the following two conditions holds: (1) $\psi(v) \neq l_1$, or (2) $\psi(v) = l_1$ and v has at least k neighbors with type l_2 in R.

The r-com model is generalized from the classic k-core on the homogeneous network. A homogeneous network can be seen as an HIN with only one vertex type, say l_0. In this case, the relational constraint $\langle l_0, l_0, k \rangle$ requires that each vertex has at least k neighbors, so such an r-com is exactly a k-core.

Since the r-com model is generalized from classic k-core, to enumerate all the maximal r-coms of a set S of relational constraint, a natural idea is to first find a subgraph induced by vertices with types in L_S, and then iteratively remove vertices that are not qualified. To speed up this process, the idea of message-passing is adopted. Specifically, it scans all the vertices and removes unqualified vertices. When removing an unqualified vertex v, each of v's neighbors will receive a message that v has been removed, and unqualified neighbors will be collected into a queue Q. All the vertices of Q will be removed in a similar manner until Q is empty. As shown in [92], this optimized algorithm only takes near-linear time cost if $|S|$ is smaller than the average degree of $\mathcal{H}$. In addition, Jian et al. [92] studied the problem of finding the r-com with the minimum size and proposed efficient approximate solutions.

4.2.3 The h-Structure Model

This model, introduced by Wang et al. [172], is a core-based cohesive subgraph on the weighted HIN with a star schema. The h-structure is a sub-HIN extracted based on vertices' s-degree by following the classic concept of h-index [80]. A weighted HIN, denoted by $\mathcal{H} = (V, E, W)$, follows the definition of HIN (see Definition 2.1), except that each edge $(v, u) \in E$ has a weight $w(u, v) \in W$. A star schema is a schema, whose network structure is a star, where the center vertex type, denoted by X_0, is called the *base-type* and other vertex types are called *attribute-types*.

The h-structure of a weighted HIN is extracted based on the s-degrees of vertices with base-types. The s-degree of a vertex v with base-type is defined by aggregating the weights between x and all its neighbors, which are grouped as base-weight and attribute-weight. Let $N(v, X)$ denote the set of neighbors of the vertex v, which are with the type X. These weights are defined as follows.

Definition 4.10 (Base-Weight and Attribute-Weight) Given a weighted HIN $\mathcal{H} = (V, E, W)$ and a vertex $v \in \mathcal{H}$, the base-weight of v is defined as

$$w(v, X_0) = \sum_{u \in N(v, X)} w(v, u), \tag{4.1}$$

and the attribute-weight of v with an attribute-type X is defined as $w(v, X) = \frac{1}{N(v, X)} \sum_{u \in N(v, X)} w(v, u)$.

Definition 4.11 (s-Degree) Given a weighted HIN $\mathcal{H}$, the s-degree of a vertex v with base-type is

$$d_s(v) = \sum_{X \in \mathcal{X}} z(w(v, X)) = \sum_{X \in \mathcal{X}} z\left(\frac{w(v, X) - \mu_\mathcal{X}}{\delta_\mathcal{X}}\right), \tag{4.2}$$

where $\mathcal{X}$ contains all the vertex types of vertices connected to v, $\mu_\mathcal{X}$ is the average value of the weights, and $\delta_\mathcal{X}$ is the corresponding standard deviation (note that $\mathcal{X}$ may contain X_0).

Figure 4.3 gives an example of computing the s-degree of a vertex. In the weighted HIN $\mathcal{H}$, the sub-network which consists of vertices with type X_0 and edges among these vertices, denoted by G_0, is a homogeneous network, also called the base-homogeneous-network of $\mathcal{H}$. After computing the s-degree of each vertex in G_0, a core structure can be located from G_0 by using the h-index [80], which is a subgraph formed by h vertices whose s-degrees are at least h. By reducing all adjacent edges to the core structure, a heterogeneous core structure, called h-structure, of the whole weighted HIN is obtained. Consequently, the h-structure of a weighted HIN is a sub-HIN that is formed by a set of base-type vertices of the highest s-degrees and edges induced to these vertices.

The h-structure of a weighted HIN can be easily extracted by first computing the s-degree of each base-type vertex, then locating the core structure of G_0 using

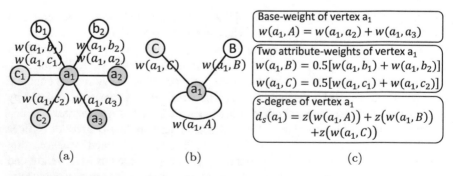

(a)   (b)   (c)

Fig. 4.3 An example of computing s-degree [172] (the base-type is "**a**", and attribute-types are "**b**" and "**c**"). (a) An HIN. (b) Type weights. (c) s-degree of a_1

h-index, and finally inducing the sub-HIN containing the core structure, which only take time cost linear to the size of the HIN.

4.2.4 The $(a_1, \cdots, a_k)$-Core Model

This core model [195] was developed on the *k*-partite network:

Definition 4.12 (($a_1, \cdots, a_k$)-Core) Given a *k*-partite network $\mathcal{H} = (V = (V_1, V_2, \cdots, V_k), E)$, a subgraph $\mathcal{H}' = (V' = (V_1', V_2', \cdots, V_k'), E')$ of $\mathcal{H}$ is a $(a_1, \cdots, a_k)$-core, if for each vertex $v \in V_i'$, it is connected to at least a_j vertices in V_j' for all $j \in [1, k]$ with $i \neq j$.

Definition 4.13 (Maximal ($a_1, \cdots, a_k$)-Core) Given a *k*-partite network $\mathcal{H}$, a $(a_1, \cdots, a_k)$-core of $\mathcal{H}$ is a maximal $(a_1, \cdots, a_k)$-core, if it is not contained in a larger $(a_1, \cdots, a_k)$-core.

For example, in the 3-partite network of Fig. 4.4, it is easy to see that the whole network is the maximal $(1, 1, 1)$-core, while the subnetwork of $\{a_1, a_2, b_2, c_1, c_2\}$ is the maximal $(1, 2, 1)$-core.

Given a *k*-partite network and a set of values for $a_1, \cdots, a_k$, Zhou et al. [195] proposed an algorithm to compute the $(a_1, \cdots, a_k)$-core, which follows the paradigm of computing the *k*-core on a unipartite network by peeling vertices one by one [14]. Specifically, for each vertex, it first computes the number of incident edges to vertices in each partite set. Then, for each vertex in each partite set V_i, it iteratively removes vertices that are without a_j edges to vertices in the partite set V_j and updates the numbers of incident edges to its neighbors. Finally, the core is computed.

4.2.5 The Multi-Layer Core Model

In Zhu et al. [203] and Liu et al. [113] respectively proposed a multi-layer core model on the multi-layer network, which were originally called multi-layer *k*-core and *k*-coherent core (*k*-CC), respectively. Here, we simply call them the *k*-CC model.

Fig. 4.4 A 3-partite network $(V_1 = \{a_1, a_2\}, V_2 = \{b_1, b_2, b_3\},$ and $V_3 = \{c_1, c_2\})$

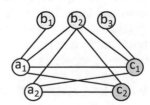

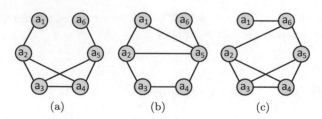

Fig. 4.5 A 3-layer network. (**a**) The 1st layer. (**b**) The 2nd layer. (**c**) The 3rd layer

Definition 4.14 (k-CC [113, 203]) Given a multi-layer network $\mathcal{H} = (V, E = (E_1, E_2, \cdots, E_l))$, a set of layer networks $\mathcal{L}$ in $\mathcal{H}$, and an integer k, the k-CC of $\mathcal{H}$ on $\mathcal{L}$ is the maximum set of vertices, such that for each $G_i = (V, E_i) \in \mathcal{L}$, they form a k-core in the layer network G_i.

Figure 4.5 shows a 3-layer network $\mathcal{H}$. If $\mathcal{L} = \mathcal{H}$, then the vertex set $\{a_2, a_3, a_4, a_5\}$ is a 2-CC of $\mathcal{H}$ on $\mathcal{L}$. Similar to the k-core computation [14], the k-CC can be computed by iteratively peeling vertices that do not satisfy the minimum degree constraint. The computation of all the possible k-CCs in a multi-layer network, also known as CoreCube [113], can be computed similarly.

Based on the k-CC model, Zhu et al. [203] studied the problem of finding top-h diversified coherent cores. Specifically, given a multi-layer network $\mathcal{H}$ and three integers k, s ($s \leq l$), and h, it aims to find a set $\mathcal{S} = \{S_1, S_2, \cdots, S_h\}$, such that (1) each S_i is the k-CC of a multi-layer network $\mathcal{H}'$ that is formed by s layers of $\mathcal{H}$; and (2) the coverage of $\mathcal{S}$, i.e., $\cup_{i \in [1,h]} S_i$, is maximized. Essentially, it finds h sets of vertices such that each set of vertices is densely connected regarding a subset of layers of $\mathcal{H}$, and meanwhile the diversity of all these h sets of vertices is maximized. For example, consider the 3-layer network in Fig. 4.5. By setting $k = 2$, $s = 2$, and $h = 1$, the vertex set $\{a_2, a_3, a_4, a_5\}$ is the top-1 diversified 2-CC, since it has the maximum coverage (i.e., it covers all the six vertices) among all the three combinations, each of which contains two layers of $\mathcal{H}$.

The problem of finding top-k diversified coherent cores in a multi-layer network is NP-hard [203]. To speed up the search by trading accuracy, Zhu et al. developed a simple greedy algorithm with an approximation ratio of $1 - \frac{1}{e} \approx 0.632$. Specifically, it enumerates all the s-layer networks that are formed by s layers of $\mathcal{H}$, computes the k-CC of each s-layer network, and returns the h k-CCs which contribute the largest increase of coverage one by one. However, as aforementioned, with the increase of l, the number of s-layer networks grows exponentially, which renders this algorithm impractical, especially when l is large. To alleviate this issue, the authors further improved the efficiency by trading accuracy and proposed two approximation algorithms, which work in bottom-up and top-down manners respectively. These two algorithms run much faster than the greedy algorithm above, but achieve a lower theoretical approximation ratio of $\frac{1}{4}$ [203, 204].

In the k-CC core model above, the set of vertices in the core must form a k-core in each layer network. This constraint can be further relaxed such that for different layers, the thresholds of the minimum degree constraint are different, which results in another more general multi-layer core model proposed by Galimberti et al. [65, 66]:

Definition 4.15 (Multi-Layer k-Core [65]) Given a multi-layer network $\mathcal{H} = (V, E = (E_1, E_2, \cdots, E_l))$ and an one-dimensional vector $\mathbf{k} = (k_1, k_2, \cdots, k_l)$, the multi-layer $\mathbf{k}$-core of $\mathcal{H}$ is the maximal multi-layer sub-network $\mathcal{H}' = (V', E' = (E_1', E_2', \cdots, E_l'))$ of $\mathcal{H}$, such that $\forall i \in \{1, 2, \cdots, l\}$, the set V' of vertices forms a k_i-core in the layer network $G_i' = (V', E_i')$.

Clearly, if $\mathbf{k} = (k, k, \cdots, k)$, then the corresponding multi-layer $\mathbf{k}$-core is equivalent to the k-CC with $\mathcal{L} = \mathcal{H}$. Similar to the classic k-core, the multi-layer $\mathbf{k}$-core can be computed by peeling.

In addition, since the number of threshold values is much larger than that of the k-CC model, many multi-layer $\mathbf{k}$-cores may share the same set of vertices. For example, in the 3-layer network of Fig. 4.5, if we let $\mathbf{k}_1 = (2, 2, 2)$ and $\mathbf{k}_2 = (2, 1, 2)$, then the multi-layer $\mathbf{k}_1$-core and $\mathbf{k}_2$-core share the vertex set $\{a_2, a_3, a_4, a_5\}$. To avoid the redundancy, Galimberti et al. [65, 66] studied the problem of enumerating all the distinct multi-layer $\mathbf{k}$-cores and also developed efficient solutions.

4.3 Truss-Based CSMs and Solutions

In the literature, there are three truss-based CSMs on HINs, namely $(k, \mathcal{P})$-Btruss, $(k, \mathcal{P})$-Ctruss [180], and $(b_1, \cdots, b_k)$-truss [195]. The first two models focus on finding a set of vertices with a specific type, while the third one is a subgraph of vertices with multiple vertex types.

4.3.1 The (k, P)-Btruss and (k, P)-Ctruss Models

These two models are formulated based on the concepts of b-triangle and c-triangle, extended from the classic triangle on homogeneous graphs by incorporating a symmetric meta-path $\mathcal{P}$.

Definition 4.16 (b-Triangle) Given an HIN $\mathcal{H}$ and a symmetric meta-path $\mathcal{P}$, we say three vertices (u, v, w) form a b-triangle, if any two vertices of them form a $\mathcal{P}$-pair.

Definition 4.17 (b-Support) Given an HIN $\mathcal{H}$, a symmetric meta-path $\mathcal{P}$, and a set S of $\mathcal{P}$-pairs, the *b-support* of a $\mathcal{P}$-pair (u, v) regarding S is the number of b-triangles, which are formed by $\mathcal{P}$-pairs in S, containing it.

Fig. 4.6 Illustrating the
b-triangle and c-triangle. (**a**)
A star. (**b**) A circle

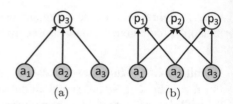

(a)   (b)

Definition 4.18 ($(k, \mathcal{P})$-Btruss) Given an HIN $\mathcal{H}$, an integer k, and a symmetric meta-path $\mathcal{P}$, the $(k, \mathcal{P})$-Btruss is the maximum set of $\mathcal{P}$-pairs, denoted by Φ_b, such that for each $\mathcal{P}$-pair, its b-support regarding Φ_b is at least $(k-2)$.

Although the model above is straightforward, it may lead to weak cohesiveness. Recall that in a homogeneous network, three vertices are said to form a triangle, if any two of them are connected by an edge. As a result, a triangle can be regarded as a 3-circle, or 3-clique, which implies that if any vertex or edge is removed, the residual part is still connected, making the cohesiveness among the three vertices strong. However, the b-triangle may not be able to achieve this since its three vertices may form a star if the three instances of $\mathcal{P}$ share the same middle vertex. For example, in Fig. 4.6, a_1, a_2, a_3 form b-triangles in stars and cycles by using a meta-path $\mathcal{P} = (APA)$. Removing the middle vertex or any edge of the star will make it disconnected. To tackle this issue, the concept of c-triangle (also called bi-triangle [181]) is proposed, which is a circle as shown in Fig. 4.6b.

Definition 4.19 (c-Triangle) Given an HIN $\mathcal{H}$ and a symmetric meta-path $\mathcal{P}$, three vertices (u, v, w) form a c-triangle, if there exist three instances of $\mathcal{P}$, such that each connects a pair of vertices and they do not share any vertex except u, v, and w.

Definition 4.20 (c-Support) Given an HIN $\mathcal{H}$, a symmetric meta-path $\mathcal{P}$, and a set S of $\mathcal{P}$-pairs, the *c-support* of a $\mathcal{P}$-pair (u, v) regarding S is the number of c-triangles, which are formed by $\mathcal{P}$-pairs in S, containing it.

Definition 4.21 ($(k, \mathcal{P})$-Ctruss) Given an HIN $\mathcal{H}$, an integer k, and a symmetric meta-path $\mathcal{P}$, the $(k, \mathcal{P})$-Ctruss is the maximum set of $\mathcal{P}$-pairs, denoted by Φ_c, such that for each $\mathcal{P}$-pair, its c-support regarding Φ_c is at least $(k-2)$.

The $(k, \mathcal{P})$-Btruss of $\mathcal{H}$ is equivalent to the classic k-truss in the $\mathcal{P}$-graph, so a simple approach of computing a $(k, \mathcal{P})$-Btruss is to first build a $\mathcal{P}$-graph and then compute the k-truss of $\mathcal{P}$-graph [164] as the result. However, the $\mathcal{P}$-graph is often much denser than $\mathcal{H}$, especially when $\mathcal{P}$ is very long, so another faster approach is to extract the truss from the HIN directly, by iteratively peeling $\mathcal{P}$-pairs whose b-supports are less than $(k-2)$ [180].

For the $(k, \mathcal{P})$-Ctruss, its computation is more challenging, since compared to a b-triangle, a c-triangle has an additional "circle" constraint, which makes it cannot be computed from the $\mathcal{P}$-graph directly because the $\mathcal{P}$-graph does not reveal the "circle" constraint. To efficiently verify whether three vertices form a c-triangle, Yang et al. [180] present a bidirectional search method, which searches the circle from the three vertices simultaneously, making the search space much smaller.

Based on the c-triangle verification algorithm, the $(k, \mathcal{P})$-Ctruss can be computed by iteratively peeling $\mathcal{P}$-pairs that do not participate in at least $(k-2)$ c-triangles.

4.3.2 The $(b_1, \cdots, b_k)$-Truss Model

This model was designed for the k-partite network with $k \geq 3$. Similar to the classic k-truss model, it relies on a novel triangle concept which we term 3-partite-triangle, by involving three vertices from three partite sets of the k-partite network respectively.

Definition 4.22 (3-Partite-Triangle) Given a k-partite network $\mathcal{H} = (V = (V_1, V_2, \cdots, V_k), E)$, a 3-partite-triangle $\triangle_{u,v,w}$ is a cycle formed by three vertices from three partite sets and three edges linking them, i.e., $u \in V_x$, $v \in V_y$, and $w \in V_z$ with $x, y, z \in [1, k]$ and $x \neq y \neq z$.

Definition 4.23 ($(b_1, \cdots, b_k)$-Truss) Given a k-partite network $\mathcal{H} = (V = (V_1, V_2, \cdots, V_k), E)$, a subgraph $\mathcal{H}' = (V' = (V'_1, V'_2, \cdots, V'_k), E')$ of $\mathcal{H}$ is a $(b_1, \cdots, b_k)$-truss, if for each edge $e = (v_i, v_j)$ with $v_i \in V_i$ and $v_j \in V_j$, its support $sup(e, \mathcal{H}') \geq \min\{(b_i - 2), (b_j - 2)\}$, where $sup(e, \mathcal{H}')$ denotes the number of 3-partite-triangles containing e in $\mathcal{H}'$.

Definition 4.24 (Maximal $(b_1, \cdots, b_k)$-Truss) Given a k-partite network $\mathcal{H}$, a $(b_1, \cdots, b_k)$-truss of $\mathcal{H}$ is a maximal $(b_1, \cdots, b_k)$-truss, if it is not contained in a larger $(b_1, \cdots, b_k)$-truss.

For example, in the 3-partite network of Fig. 4.4, the edge $e_1 = (a_1, b_1)$ does not participate in any 3-partite-triangle, while the edge $e_2 = (a_1, b_2)$ involves into two 3-partite-triangles, which are formed by vertex sets $\{a_1, b_2, c_1\}$ and $\{a_1, b_2, c_2\}$ respectively, so $sup(e_1, \mathcal{H}) = 0$ and $sup(e_2, \mathcal{H}) = 2$. Besides, it is easy to observe that the whole network is the maximal $(2, 2, 2)$-truss, while the subnetwork of $\{a_1, a_2, b_2, c_1, c_2\}$ is the maximal $(4, 4, 3)$-truss.

Given a k-partite network and a set of values for $b_1, \cdots, b_k$, Zhou et al. [195] proposed an efficient algorithm to compute the $(b_1, \cdots, b_k)$-truss, which generally follows the paradigm of computing the classic k-truss on a unipartite network by peeling edges one by one [164]. Specifically, first, for each edge e, it computes e's support $sup(e, \mathcal{H})$, i.e., the number of 3-partite-triangles that contains it. Then, it uses a loop and in each iteration of the loop, it removes an edge e with $sup(e, \mathcal{H}') < \min\{(b_i - 2), (b_j - 2)\}$ and updates the support values of its connected edges. Finally, the maximal $(b_1, \cdots, b_k)$-truss is computed.

4.4   Clique-Based CSMs and Solutions

In this section, we review the clique-based CSMs on HINs, which include the maximal motif-clique [81], ABCOutlier-clique [76], cliques on k-partite graphs [47, 129, 139], and cliques on multi-layer graphs [19, 138, 183].

4.4.1   The Maximal Motif-Clique Model

In [81], Hu et al. introduced the maximal motif-clique (m-clique for short) on undirected HINs based on a user-given motif (a.k.a. higher-order structure or graphlet). A motif of an HIN $\mathcal{H}$ is a small connected graph $M = (V_M, E_M)$, which follows the schema of $\mathcal{H}$, i.e., (1) only contains vertex types defined by the schema; and (2) only contains edges between vertex types that are allowed by the schema. Essentially, a motif describes the composite relationship among a set of vertex types, and it serves as a fundamental building block of large complex HINs. The m-clique is formally defined based on the concepts of subgraph isomorphism and type-matched vertex set.

Definition 4.25 (Subgraph Isomorphism) A motif M is subgraph isomorphic to an HIN $\mathcal{H} = (V, E)$ if there exists an injective mapping $\xi : V_M \rightarrow V$, s.t., $\forall u \in V_M$, $\psi(u) = \psi(\xi(u))$, and $\forall u, v \in V_M$, if $(u, v) \in E_M$, then $(\xi(u), \xi(v)) \in E$, where $\xi(u)$ is the vertex to which u is mapped in $\mathcal{H}$.

Definition 4.26 (Type-Matched Vertex Set) Given an HIN $\mathcal{H} = (V, E)$ and a motif $M = (V_M, E_M)$, a vertex set $S \subseteq V$ is a type-matched vertex set of M, if S and M have the same number of vertices and there exists a bijection $\zeta : S \rightarrow V_M$, s.t., $\forall u \in S$, $\psi(u) = \psi(\zeta(u))$.

Definition 4.27 (m-Clique) Given an HIN $\mathcal{H} = (V, E)$ and a motif $M = (V_M, E_M)$, an m-clique of M in $\mathcal{H}$ is an induced HIN $\mathcal{H}'$ of $\mathcal{H}$, s.t., $\mathcal{H}'$ and M have the same set of vertex types and for each type-matched vertex set S in $\mathcal{H}'$, M is subgraph isomorphic to the induced subgraph of S in $\mathcal{H}'$.

Definition 4.28 (Maximal m-Clique) Given an HIN $\mathcal{H} = (V, E)$ and a motif $M = (V_M, E_M)$, an m-clique is a maximal m-clique if it is not contained in any other m-clique.

Figure 4.7 shows an example undirected HIN $\mathcal{H}$ following the DBLP network schema, and a triangle motif M that describes an author having two papers with citation relationship. Clearly, there are two maximal m-cliques of M, as depicted in Fig. 4.7c,d respectively. More real examples can be found in the demonstration system [101]. Note that when the HIN is a k-partite network ($k \geq 2$), the maximal m-clique is equivalent to the maximal k-partite clique (see Sect. 4.4.3).

The problem of discovering the maximal m-cliques has been proved to be NP-hard. To discover the maximal m-cliques, Hu et al. [81] proposed an efficient

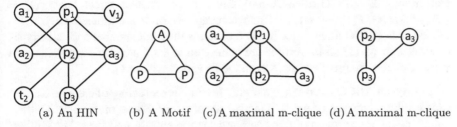

(a) An HIN   (b) A Motif   (c) A maximal m-clique   (d) A maximal m-clique

Fig. 4.7 Illustrating the maximal m-cliques [81]. (**a**) An HIN. (**b**) A Motif. (**c**) A maximal m-clique. (**d**) A maximal m-clique

algorithm META by following the BK algorithm [21], which is the classic algorithm for maximal clique enumeration. Given an unlabelled graph, BK uses backtracking to recursively search all maximal cliques. The intuition is to expand an existing clique to a larger one iteratively. It maintains three disjoint sets: a set U of vertices in the current detected clique, a set C of candidates for clique expansion (i.e., $\forall v \in C$, $U \cup \{v\}$ form a clique), and a set NOT of forbidden vertices (i.e., $\forall v \in NOT$, $U \cup \{v\}$ form a clique, and all maximal cliques containing $U \cup \{v\}$ have already been found). Note that NOT ensures every maximal clique will only be reported once. Initially, both U and NOT are empty and all the vertices are in C. Then, the algorithm iteratively adds each candidate $u \in C$ to U, updates C and NOT to include only neighbors of u, and recursively invokes BK to find all maximal cliques containing $U \cup \{u\}$. A maximal clique is reported when both C and NOT are empty. The above procedure is repeated until all maximal cliques are found.

Following the idea of BK, META has the following optimizations: (1) it adopts a state-of-the-art subgraph isomorphism algorithm to find all embeddings of M, which are used to initialize the set C; (2) it optimizes the process of vertex expansion, i.e., checking whether a vertex can be incorporated into an m-clique to become a larger one; (3) it includes a novel early stop technique, which can stop expanding early; and (4) it uses the set-trie tree to avoid computing duplicate maximal m-cliques.

4.4.2   The ABCOutlier-Clique Model

In [76], Gupta et al. detected the association-based clique outlier (ABCOutlier in short) from attributed HINs and modeled outliers by introducing a clique-based subgraph model which we term ABCOutlier-clique. In the attributed HIN, each vertex type $A_i \in \mathcal{A}$ is associated with a set of D_i attributes $(A_{i1}, A_{i2}, \cdots, A_{iD_i})$. To detect ABCOutliers, a conjunctive select query Q of L (vertex type, predicate) pairs is required, i.e., $Q = \langle (A_1, P_1), (A_2, P_2), \cdots, (A_L, P_L) \rangle$, where A_i is a vertex type and P_i is a predicate defined only on the attributes of type A_i.

Definition 4.29 (ABCOutlier-Clique) Given a conjunctive select query $Q = \langle (A_1, P_1), (A_2, P_2), \cdots, (A_L, P_L) \rangle$, a subgraph C of the attributed HIN $\mathcal{H} = (V, E)$ is an ABCOutlier-clique if: (1) it contains the same number and same types of entities in Q; (2) all its vertices are connected to each other in $\mathcal{H}$; and (3) each vertex satisfies the predicate defined on the its vertex type in Q.

Clearly, an ABCOutlier-clique not only has the link structure of a clique but also satisfies the predicates of the conjunctive select query. Based on the ABCOutlier-clique model, the problem of ABCOutlier detection aims to find the k ABCOutlier-cliques with the highest outlier scores. The outlier score of an ABCOutlier-clique C is defined as follows.

$$score(C) = \sum_{e=(T_i, T_j) \in C} \frac{\sum_{a=1}^{D_i} \sum_{b=1}^{D_j} score(v_a, v_b)}{D_i \times D_j} \quad (4.3)$$

Here $score(v_a, v_b)$ is the score between the a-th and b-th attribute values of types T_i and T_j respectively, which considers the co-occurrence frequency and peakedness of v_a and v_b [76].

To find the ABCOutlier-cliques, a naive method is to perform an exhaustive search of all possible one-to-one correspondences of query pairs in Q to vertices in the HIN of the matched type. Obviously, this method will incur a complexity of exponential time cost. To facilitate efficient query, Gupta et al. [76] proposed an index-based approach, which consists of an offline phase and an online phase. In the offline phase, an index structure, which stores some shared neighbor information for each vertex, is built. In the online phase, when a query Q comes, it first decomposes Q into a list of query pairs, then finds the matched vertex pairs for each query pair, and finally computes the ABCOutlier-cliques by using the matched vertex pairs of these query pairs, during which the index is used to prune vertices that are not in the ABCOutlier-cliques. Based on the ABCOutlier-cliques enumeration algorithms, the top-k ABCOutliers can be extracted by computing their scores. The details [76] of scoring and ranking are omitted due to the space limitation.

4.4.3 The k-Partite Clique Model

This model (k-PC for short) was developed for the k-partite network.

Definition 4.30 (k-PC [115, 139]) Given a k-partite network $\mathcal{H} = (V = (V_1, V_2, \cdots, V_k), E)$, a k-PC is a subnetwork $\mathcal{C} = (V' = (V_1', V_2', \cdots, V_k'), E')$ of $\mathcal{H}$, which satisfies that (1) $\forall i \in [1, k]$, $V_i' \subseteq V_i$ and $E' \subseteq E$; and (2) $\forall u \in V_i', v \in V_j'$ with $i \neq j$, they are linked by an edge (called intra-partite edge).

Definition 4.31 (Maximum k-PC [115, 139]) Given a k-partite network $\mathcal{H} = (V = (V_1, V_2, \cdots, V_k), E)$, a k-PC $\mathcal{C}$ is the maximum k-PC, if there is no other k-PCs that are larger than $\mathcal{C}$.

Definition 4.32 (Maximal k-PC [115, 139]) Given a k-partite network $\mathcal{H} = (V = (V_1, V_2, \cdots, V_k), E)$, a k-PC $\mathcal{C}$ is a maximal k-PC, if no vertex can be added into $\mathcal{C}$ to form a larger k-PC.

Essentially, a k-PC is a complete k-partite network that contains all possible inter-partite edges among all the k parties. The maximum and maximal k-PCs are the k-PCs, which have the largest sizes and achieves the structure maximality, respectively. For example, in the 3-partite network of Fig. 4.4, it is easy to observe that vertices $\{a_1, b_2, c_1\}$ form a 3-PC, while vertices $\{a_1, a_2, b_2, c_1, c_2\}$ form the maximum 3-PC which is also the maximal 3-PC.

To enumerate the maximum and maximal k-PCs of a k-partite network, several approaches have been developed. The state-of-the-art approach, called MMCE (or maximal multipartite clique enumeration), was proposed by Phillips et al. [139]. MMCE is based on the BK algorithm [21], which is a classic maximal k-clique enumeration algorithm for the unipartite network, as discussed in Sect. 4.4.1. Specifically, MMCE first adds all the vertices in the k partite sets and intra-partite edges to its input. Then, it initializes three BK-style vertex sets (see Sect. 4.4.1). After that, it invokes a function, which is a recursive BK-style subroutine modified to check whether a maximal clique contains a vertex from each partite set. Finally, all the maximal k-PCs are enumerated and outputted as they are discovered, which implies they do not need to be stored, resulting in little space cost.

Besides, there are several other similar definitions of k-PCs in the literature, which can be considered as variants of the k-PC model:

(1) Grunert et al. [74, 129] developed a simplified version of the k-PC, which only requires one vertex from each of the k partite sets, making the total number of vertices in a k-PC be exactly k. To enumerate such k-PCs in a k-partite network, [74, 129] present some efficient k-PC enumeration algorithms, which adopt the branch-and-bound paradigm.

(2) In [47], Dawande et al. focused on a simplified network model of the k-partite network, which is also represented by $\mathcal{H} = (V = (V_1, V_2, \cdots, V_k), E)$, but it can considered as a network consisting of k levels, where each level corresponds to a partite set, and requiring that for each edge $(u, v) \in E$, vertices u and v are from two adjacent levels, i.e., $u \in V_i$ and $v \in V_{i+1}$. Based on this data model, they introduced another multipartite clique, denoted by $M = U_1 \cup U_2 \cup \cdots \cup U_m$, which satisfies that (1) $\forall i \in [1, m]$ with $m \geq 2$, $U_i \subseteq V_{j+i}$ with some $j \geq 0$; and (2) $\forall u_1 \in U_i, u_2 \in U_{i+1}, (u_1, u_2) \in E$.

(3) In [195], Zhou et al. proposed a relaxed version of the k-PC model, called k^*-PC. Instead of strictly requiring vertices from all the k partite sets, this model aims to find k'-PCs from a subnetwork of the k-partite network $\mathcal{H}$, which is induced by vertices of only k' partite sets in $\mathcal{H}$, where $k' \leq k$. They focused on finding the edge-maximum k^*-PC, which is the k-PC with the maximum number of edges, and proposed efficient enumeration algorithms.

4.4.4   Multi-Layer Quasi-Clique Models

The first quasi-clique model on multi-layer networks, called cross-graph quasi-clique, was introduced by Pei et al. [138], which is defined based on the concept of γ-quasi-complete graph on the homogeneous network.

Definition 4.33 (γ-Quasi-Complete Graph) Given a homogeneous network $G = (V, E)$ and a real value $\gamma \in (0, 1]$, the connected subgraph induced by a set S of vertices in G is a γ-quasi-complete graph if every vertex of S has a degree at least $\gamma \cdot (|S| - 1)$ within the subgraph.

Definition 4.34 (Cross-Graph Quasi-Clique) Given a multi-layer network $\mathcal{H} = (V, E = (E_1, E_2, \cdots, E_l))$ and real values $\gamma_1, \gamma_2, \cdots, \gamma_l$ ($i \in [1, l]$ and $\gamma_i \in (0, 1]$), a set S of vertices in $\mathcal{H}$ is a cross-graph quasi-clique, if (1) in each layer graph $G_i = (V, E_i)$, the subgraph induced by S is a γ_i-quasi-complete graph; and (2) there is no proper superset of S has this property.

Clearly, when $\gamma_i = 1$ for all $i \in [1, l]$, then the cross-graph quasi-clique forms a clique in each layer graph. For example, consider the 3-layer network in Fig. 4.5. If $\gamma_1 = \gamma_2 = \gamma_3 = 0.6$, then the set of vertices $S = \{a_1, a_2, a_3, a_4\}$ forms a cross-graph quasi-clique, since in each layer graph, every vertex's degree is 2, which is larger than $0.6 \times (4 - 1) = 1.8$.

Given a multi-layer network, the problem of enumerating its complete set of cross-graph quasi-cliques is NP-hard. To solve this problem, Pei et al. developed an efficient algorithm, called Crochet. It relies on a key idea that given a set S of vertices and a total order on S, the complete set of various subsets of S (i.e., $2^{|S|}$) can be enumerated systematically using a set enumeration tree, where some nodes represent the cross-graph quasi-cliques. Crochet conducts a depth-first search on the set enumeration tree of vertices to find the cross-graph quasi-cliques. Meanwhile, some effective pruning criteria are developed to improve the efficiency [138]. In addition, Zeng et al. [183] studied an important variant of the cross-graph quasi-clique, called the coherent closed quasi-clique, and investigated a more general problem of mining the frequent coherent closed quasi-cliques.

Besides, in [19, 20], Boden et al. extended the above models for finding coherent subgraphs from multi-layer networks with edge labels, and proposed the multi-layer coherent subgraph (MLCS) model. A labelled multi-layer network $\mathcal{H} = (V, E_1, L_1, \cdots, E_l, L_l)$ consists of l edge-labelled layers with each denoted by $G_i = (V, E_i, L_i)$. Specifically, they first introduced the model of one-dimensional MLCS cluster on a single layer network and then extended it for multi-layer networks.

Definition 4.35 (One-Dimensional MLCS Cluster) Given a labelled network layer $G_i = (V, E_i, L_i)$, a distance function $dist$ defined on edge labels, and a threshold w, a set $S \subseteq V$ of vertices is a one-dimensional MLCS cluster if its induced subgraph in G_i is a 0.5-quasi-complete graph, and for any two edges e_1, e_2 formed by vertices in S, $dist(L_i(e_1), L_i(e_2)) \leq w$ holds.

Definition 4.36 (MLCS Cluster) Given a labelled multi-layer network $\mathcal{H} = (V, E = (E_1, L_1, \cdots, E_l, L_l))$, a distance function $dist$, and a threshold w, a set $S \subseteq V$ of vertices regarding a subset of layers R is an MLCS cluster, if $\forall i \in R$, S is a one-dimensional MLCS in the layer network $G_i = (V, E_i, L_i)$.

Note that when $\mathcal{H}$ is unlabelled, the function $dist$ is not applicable and the MLCS model resembles the definitions of cross-graph quasi-cliques and coherent closed quasi-cliques above. Enumerating all the MLCS clusters may lead to a large number of valid clusters that contain much overlap redundant information. To avoid redundancy, Boden et al. introduced the redundancy relation and then proposed to select the maximum-quality results that should not contain redundant clusters. This problem is NP-hard and they developed an efficient approximate algorithm in [19, 20].

4.5   Density-Based CSMs and Solutions

In the literature, the works of density-based cohesive subgraphs mainly focus on the multi-layer network. The density of a multi-layer network was first introduced by Jethava et al. [91], and it is also called *common density*, which is extended from the classic edge density in Definition 2.9.

Definition 4.37 (Common Density) Given a multi-layer network $\mathcal{H} = (V, E = (E_1, E_2, \cdots, E_l))$ and a set S of vertices in $\mathcal{H}$, the common density of S is

$$\rho(\mathcal{H}, S) = \min_{i \in \{1, \cdots, l\}} \rho(G_i, S) = \min_{i \in \{1, \cdots, l\}} \frac{E_i(S)}{|S|}, \quad (4.4)$$

where $|E_i(S)|$ denotes the number of edges connecting vertices of S in the layer network G_i.

Definition 4.38 (Common Densest Subgraph (CDS)) Given a multi-layer network $\mathcal{H} = (V, E = (E_1, E_2, \cdots, E_l))$, the CDS is the set S^* of vertices satisfying

$$S^* = arg \max_{S \subseteq V} \rho(\mathcal{H}, S). \quad (4.5)$$

Clearly, the common density of a set S of vertices is exactly the minimum edge density of S among all the layers of $\mathcal{H}$, and the CDS is the set of vertices with the highest common density. To discover the CDS, Jethava et al. presented a greedy approximation algorithm. The algorithm first initializes a set $V_0 = V$ and a timestamp $t = 0$, and then iteratively constructs a vertex set V_{t+1} at each time t by removing the vertex v_t from V_t where v_t is the minimum degree vertex in the subgraph $G(V_t)$, which has the minimum density among all the subgraphs induced by V_t in $\mathcal{H}$. Note, however, this algorithm has no provable theoretical guarantee of the approximation ratio.

A major limitation of CDS is that considering all layers, even the noisy or insignificant layers would contribute to selecting the output subgraph, which would be not really dense, thus ignoring the subgraphs which are dense but still in a large subset of layers. To overcome this limitation, Galimberti et al. [65, 66] generalized the common density, and proposed the *multi-layer density* by accounting for a trade-off between high density and number of layers exhibiting the high density.

Definition 4.39 (Multi-Layer Density) Given a multi-layer network $\mathcal{H} = (V, E = (E_1, E_2, \cdots, E_l))$, a positive real number β, and a set S of vertices in $\mathcal{H}$, the multi-layer density of S is

$$\delta(\mathcal{H}, S) = \max_{\widehat{L} \subseteq \{1, \cdots, l\}} \min_{i \in \widehat{L}} \frac{|E_i(S)|}{|S|} |\widehat{L}|^{\beta}, \tag{4.6}$$

where $|E_i(S)|$ denotes the number of edges connecting vertices of S in the layer network G_i.

Definition 4.40 (Multi-Layer Densest Subgraph (MDS)) Given a multi-layer network $\mathcal{H} = (V, E = (E_1, E_2, \cdots, E_l))$, the MDS is the set S^* of vertices satisfying

$$S^* = arg \max_{S \subseteq V} \delta(\mathcal{H}, S). \tag{4.7}$$

The role of β is to control the importance of the two ingredients of the multi-layer density, i.e., the density and number of layers exhibiting such a density. Clearly, the smaller β, the larger the importance to be given to the density, and vice versa. Note that when $\beta = 0$ and $\widehat{L} = \{1, \cdots, l\}$, the MDS is equivalent to the CDS.

To compute the MDS, Galimberti et al. developed a simple approximation algorithm based on the multi-layer **k**-core model which is discussed in Sect. 4.2.5. Specifically, it first performs multi-layer core decomposition which computes all the multi-layer **k**-cores of $\mathcal{H}$, and then among all these cores, returns the one maximizing the multi-layer density. It is proved that this algorithm achieves an approximation ratio of $\frac{1}{2l^{\beta}}$.

4.6   Other CSMs and Solutions

In [13], Barman et al. considered a special type of HIN, which can be decomposed into one primary subgraph $G_0 = (V_0, E_0)$, one or several secondary subgraphs $G_i = (V_i, E_i)$ with $i = 1, 2, \cdots$, and the bipartite networks $G_{i,j}$ among them, where each bipartite network $G_{i,j}$ is formed by vertices from G_i and G_j and edges connecting them. The edges in a subgraph $G_i (i = 0, 1, \cdots)$ are called intra-edges, while edges in bipartite networks $G_{i,j}$ are called inter-edges. Based on this HIN model, the k-context structure was proposed, which is a cohesive subgraph extracted

by analyzing the intra-edges in the primary subgraph and secondary subgraph, and
the inter-edges between them.

Definition 4.41 (k-Context Structure [13]) Given an HIN $\mathcal{H}$ that has a primary
subgraph $G_0 = (V_0, E_0)$ and a secondary subgraph $G_1 = (V_1, E_1)$, an integer k,
and a threshold θ, a set of vertices $C_i \subseteq V_0$ is a k-context structure of $v_i \in V_0$, iff
for each $v_j \in C_i$,

- There are at least two edges (v_i, v_k) and (v_j, v_l), such that v_k and v_l are from
  G_1;
- The length of the shortest path between vertices v_k and v_l is less than θ;
- The total number of such shortest paths between vertices v_i and v_j is at least k.

To make the relationship between two vertices in the k-context structure more
clearly, the authors developed a concept of *context path*, which formally establishes
their relationship via intra- and inter-edges of the secondary subgraph:

Definition 4.42 (Context Path [13]) Consider an HIN $\mathcal{H}$ which has a primary
subgraph $G_0 = (V_0, E_0)$ and a secondary subgraph $G_1 = (V_1, E_1)$, and two
vertices $v_i, v_j \in G_0$. Suppose there are two vertices $v_p, v_q \in G_1$ such that: (1)
there exist edges (v_i, v_p) and (v_q, v_j) in the bipartite network $G_{0,1}$; and (2) there
exists a path between v_p and v_q in G_1, denoted by $(v_p, \cdots, v_q)$. Then, the overall
path connecting v_i and v_j via the path above, denoted by $(v_i, v_p, \cdots, v_q, v_j)$, is a
context path.

Essentially, a k-context structure is a set of vertices in the primary subgraph, such
that each pair of its vertices is connected by at least k context paths. For example,
Fig. 4.8 shows an HIN with a primary subgraph G_0 and a secondary subgraph G_1.
Clearly, a_1 and a_2 are two vertices in G_0, and they are respectively connected to
b_1 and b_2 in G_1. Since b_1 and b_2 are connected by a path with two edges in G_1,
the path of $(a_1, b_1, b_5, b_2, a_2)$ is a context path. Besides, it is easy to verify that by
setting $k = 1$ and $\theta = 2$, the context structure of a_1 is the vertex set $\{a_1, a_2, a_3\}$.

To extract the k-context structure in an HIN with a primary subgraph and a
secondary subgraph, Barman et al. [13] proposed an algorithm, which first finds all
the pairs of vertices in the primary subgraph such that each pair of them is connected

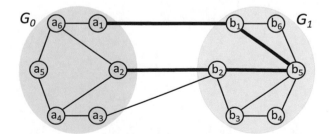

Fig. 4.8 Illustrating the k-context structure and context path

by at least k context paths, and then identifies the set of vertices such that any pair of vertices is connected by at least k context paths. Besides, the k-context structure has been extended for more complicated HINs (e.g., centralized and hierarchical HINs), which have one primary subgraph but several secondary subgraphs. The algorithm above can be easily extended for searching the k-context structure from these HINs.

4.7   Conclusions

In this chapter, we have extensively reviewed the CSMs and solutions over general HINs respectively. The core-, truss-, and clique-based models have received the most interest, while the density- and connectivity-based cohesive subgraphs have received little or no research attention. Most of these works focus on general HINs, but a few of them focus on two special types of HINs, i.e., k-partite networks and multi-layer networks. We will extensively compare and analyze these models and solutions in Chap. 5.

Chapter 5
Comparison Analysis

5.1 Bipartite Networks

As shown in Table 1.1, the CSMs on bipartite networks can be divided into five groups according to the classic CSMs, namely core-, truss-, clique-, connectivity-, and density-based models. In the following, we first make a comparison analysis for these five groups of models, and then sequentially analyze all the models in each group.

5.1.1 Comparison of Different Groups of CSMs

We first analyze the five groups of models from the following different angles:

(1) *The cohesiveness constraints are formulated differently for these groups of models.* The core-based models are usually focused on imposing constraints on each vertex, and the truss-based models often impose constraints on the edges. The clique-, connectivity- and density-based models are usually focused on the whole subgraph; that is, all the vertices in the subgraph are connected cohesively under a specific constraint defined on the whole subgraph.

(2) *In the first three groups, most of the models have the nested property, i.e., a more cohesive subgraph is contained in a less cohesive subgraph under a specific model constraint.* For instance, given a bipartite network B and integers α, β, and k, the $(\alpha + 1, \beta)$-core is contained in the (α, β)-core, and the $(k + 1)$-bitruss is contained in the k-bitruss. For core- and truss-based models, this property allows us to decompose the bipartite network under a specific model (e.g., (α, β)-core and k-bitruss) in a hierarchical manner and build the index [112] to organize these cohesive subgraphs. For the clique-based models, we aim to retrieve the maximal or maximum bicliques under specific constraints.

© The Author(s), under exclusive license to Springer Nature Switzerland AG 2022
Y. Fang et al., *Cohesive Subgraph Search Over Large Heterogeneous Information Networks*, SpringerBriefs in Computer Science,
https://doi.org/10.1007/978-3-030-97568-5_5

It is also clear to see that a biclique with larger number of vertices contains several bicliques with less number of vertices. This property allows us to develop branch-and-bound algorithms [123, 189] to retrieve these maximal or maximum bicliques. For the existing connectivity- and density-based models on bipartite networks, this nested property does not hold.

(3) *Similar to the unipartite network cases, there exists a trade-off between the computational cost and structure cohesiveness for these models.* Generally, solving the clique-based problems is the most cost-prohibitive. Specifically, the maximal biclique enumeration problem, the maximum edge biclique problem and the maximum balanced biclique problem are all NP-hard problems. Although the maximum vertex biclique problem is solvable in polynomial time, it still needs $O(nm)$ time by using the well-known Ford-Fulkerson algorithm. On the other hand, the core- and truss-based problems are easier to solve. For instance, indexes can be built on billion-scale and one-hundred-million-scale datasets to support the retrieval of the (α, β)-core vertex set and k-bitruss in optimal time, respectively [111, 169]. Note that the maximum edge biclique search problem can also be solved on billion-scale datasets, however it needs to run a totally online algorithm which is more time-consuming [123].

Considering the above trade-off, the application scenarios of different groups of CSMs can be different. The retrieved subgraphs by clique-based models is fully-connected internally and the cohesiveness of the subgraphs are usually higher that of the subgraphs obtained by the core/truss-based models. Thus, the clique-based models (e.g., maximum edge biclique) are usually used in the applications where high-effective results are needed and long-time offline computations can be afforded (e.g., fraud detection) [123]. On the other hand, the core- and truss-based models are usually preferred in the scenarios where queries are frequently posted (e.g., online recommendations) [111].

Next, we perform a more detailed comparison analysis for all the models in each of the first three groups. The other two groups are skipped as each of them has only one or two models.

5.1.2   Comparison of Different Core-Based Models

As shown in Table 5.1, we make a comprehensive comparison of these core-based models:

(1) These models have different parameters, and the concepts of degree are also formulated in different ways. In all these models, the vertices are imposed with constraints. For the first three models, since all of them rely on the (α, β)-core concept, the parameters α and β are used. In addition, the τ-strengthened (α, β)-core can be considered as a variant of the (α, β)-core which further involves constraints on edges. The generalized two-mode core can be considered as a generalization of the (α, β)-core which formulates the degree with two

Table 5.1 Core-based models and solutions on bipartite networks ("—": not applied)

Model	Model properties			Algorithm ideas	
	Parameters	Degree	Constraints	Online	Index-based
(α, β)-core [5]	α, β: min degree	#. of neighbors	Vertices	Peeling	Decomposition, nested organization
τ-strengthened (α, β)-core [79]	α, β: min degree τ: tie strength	#. of neighbors	Vertices and edges	Peeling	Decomposition, partial indexes
Generalized two-mode core [23]	α, β: min degree f, g: functions	Generated by functions	Vertices	Peeling	—
Fractional k-core [68]	k: min degree	Generated by projection	Vertices	Projection, peeling	—

     monotonic and local vertex property functions. Besides, the fractional k-core uses a graph projection approach to define the model, so it only needs a single parameter k to impose the degree constraint on the projected unipartite network.

(2) For all these models, the online computation algorithms follow the peeling paradigm. Here *"peeling"* is a paradigm which iteratively removes the vertices/edges do not meet the constraints from the network in each iteration, until no such vertex/edge remains. Note that for the fractional k-core, the graph projection approach should be used at first to generate a weighted unipartite network. The index-based methods have been proposed for the first two models. For the (α, β)-core, a complete index can be built with decomposition-based algorithms and all the vertex sets of (α, β)-cores can be retrieved in optimal time cost. Here *"decomposition"* is a method to decompose networks into a set of sub-networks following a specific CSM. Since the τ-strengthened (α, β)-core is more cohesive than the (α, β)-core, the complete index is cost-prohibitive to be built. Three partial indexes have been proposed instead with neural-network-based index selection strategies to efficiently obtain the τ-strengthened (α, β)-cores [79].

5.1.3 Comparison of Different Truss-Based Models

We compare the three truss-based models on bipartite network in Table 5.2: (1) The k-bitruss and k-tip models are directly defined on bipartite networks while the quasi-truss is based on the projected unipartite network. Therefore, k-bitruss and k-tip are based on the butterfly structure, and quasi-truss is based on the triangle structure, respectively. (2) The k-bitruss and quasi-truss models impose constraints on edges while the k-tip model involves constraints on vertices. (3) The peeling-based algorithms have been proposed to obtain these cohesive subgraphs in an

Table 5.2 Truss-based models and solutions on bipartite networks ("—": not applied)

Model	Model properties			Algorithm ideas	
	Parameters	Basic structure	Constraints	Online	Index-based
k-bitruss [206]	k: min support	Butterflies	Edges	Peeling	Decomposition, parallelization
k-tip [143]	k: min support	Butterflies	Vertices	Peeling	Decomposition, parallelization
Quasi-truss [109]	Q: min support	Triangles	Edges	Projection, peeling	Decomposition

Table 5.3 Clique-based models and solutions on bipartite networks ("—": not applied)

Model	Model properties		Optimization objective	Algorithm ideas
	Parameter-free	Completeness		
Maximal biclique [102]	Yes	Yes	Maximal subgraph	Branch-and-bound, parallelization
Maximum vertex biclique [47]	Yes	Yes	Maximum subgraph, #. of vertices	Max-flow
Maximum edge biclique [47]	Yes	Yes	Maximum subgraph, #. of edges	Branch-and-bound, progressive optimization
Maximum balanced biclique [128]	Yes	Yes	Maximum subgraph, balanced subgraph	Branch-and-bound, dynamic programming
Quasi-biclique [150]	No	No	Based	Parameter-based

online manner. The decomposition algorithms have been also proposed for all these models. By using the decomposition results, we can obtain a specific cohesive subgraph (i.e., a k-bitruss with a specific k) in optimal time cost. Note that the parallel decomposition algorithms [149] have been studied in the literature for both k-bitruss and k-tip.

5.1.3.1   Comparison of Different Clique-Based Models

As presented in Table 5.3, the clique-based models have been proposed for bipartite networks. The first four models aim to compute a maximal (or maximum) complete bi-subgraph (i.e., biclique) and the quasi-biclique models are relaxations of biclique models. Therefore, the first four models do not need input parameters while a quasi-biclique model often needs a parameter to define the error-tolerance. Note that the maximum edge biclique, maximum vertex biclique, and maximum balanced biclique models can be considered as variants of the maximal biclique model. They extend the maximal biclique model by involving constraints from different aspects to define the maximality. Besides, due to the different cohesiveness constraints, the subgraphs of the first four models are complete bipartite networks, but the quasi-biclique may not be a complete bipartite network. For algorithms to search these subgraphs, since the maximal biclique enumeration problem, the maximum edge biclique problem, and the maximum balanced biclique problem are all NP-hard problems, the proposed algorithms for solving these problems are all based on the branch-and-bound framework. In addition, the maximum vertex biclique problem can be solved in polynomial time using the max-flow-based algorithm [78]. Note that the maximum flow algorithm aims to find the flow from a source vertex to a sink vertex with the maximum flow rate in a directed weighted network.

5.2 Other General HINs

The CSMs on other general HINs can also be divided into five groups, namely core-, truss-, clique-, density-based models, and others. Next, we first make a general comparison analysis for these groups, and then analyze all the models in each group respectively.

5.2.1 Comparison of Different Groups of CSMs

To compare these five groups of models, we mainly focus on analyzing their key differences and common characteristics.

(1) *In each of the first three groups, some models consider only one target type of vertices, while others involve multiple vertex types.* More precisely, in the $(k, \mathcal{P})$-cores, multi-layer cores, $(k, \mathcal{P})$-Btruss, $(k, \mathcal{P})$-Ctruss, and quasi-cliques, their vertices are with the same type. Other models in these three groups consider vertices with multiple types. Besides, the two models in the last two groups (i.e., the multi-layer densest subgraph and k-context) only consider one target type.

(2) *In the first three groups, most of the models share the property of nested relationship*; that is, under a specific model, a more cohesive one is often nested within a less cohesive one (e.g., the $(k, \mathcal{P})$-Btruss is nested in another $(k', \mathcal{P})$-Btruss if $k \geq k'$). This property ensures that when decomposing the HIN into subgraphs following a specific model, the more cohesive subgraphs can be computed directly from the less cohesive ones, rather than the entire HIN, and these subgraphs can be organized compactly into a sequence of nested subgraphs. Note that the h-structure and multi-layer densest subgraph do not have this property as they are unique for a given network.

*(3) In line with cohesive subgraphs on unipartite networks [58], there is a trade-off
between the computational cost and structure cohesiveness for these groups of
models.* That is, the clique-based models are the most computationally costly,
but they can be used to find subgraphs that are connected very cohesively.
On the other hand, the core-based subgraphs tend to be computed easily by
following a peeling paradigm, but they may suffer from the issue of weak
cohesiveness. For example, the maximal r-com [92] can be computed in almost
linear time cost, while the problem of discovering maximal motif clique [81]
is proved to be NP-hard. For other models that are not in the groups of core-
and clique-based models, they often have higher cohesiveness than the core-
based models, but lower than the clique-based models, and meanwhile their
computational cost may be higher than that of core-based models but lower
than that of clique-based models.

*(4) For some models across different groups, there exist the inclusion-ship among
them, which may reflect their cohesiveness relationship.* For example, the
vertices of a $(k, \mathcal{P})$-Ctruss or a $(k, \mathcal{P})$-Btruss must be contained in the basic
$(k', \mathcal{P})$-core if $k \geq k' + 1$; The $(b_1, \cdots, b_k)$-truss is nested within the $(a_1, \cdots, a_k)$-core, if $b_i \geq \Sigma_{j \neq i} a_j + 1$ [196].

Next, we perform a more detailed comparison analysis for all the models in each
of the first three groups. The other two groups are skipped as each of them has only
one model.

5.2.2 Comparison of Different Core-Based Models

As shown in Table 5.4, we make a comprehensive comparison of these models:
(1) Most of the models focus on general HINs, while the last two core-based
models are developed for two special types of HINs (i.e., multi-partite and multi-
layer networks) respectively. (2) These models have different parameters, and the
concepts of degree or neighbor are also formulated in different manners. Besides,
in some core models (i.e., $(k, \mathcal{P})$-cores and multi-layer core), all the vertices are
with a single vertex type, while for others they may involve multiple vertex types,
which make the application scenarios of these models different. For example, the
$(k, \mathcal{P})$-cores are used to find communities with a specific type of vertices [62], and
the r-com is applied to finding heterogeneous communities with multiple types of
vertices [92]. (3) For most of the models, the online computation algorithms follow
the peeling paradigm, but some of them (i.e., edge- and vertex-disjoint $(k, \mathcal{P})$-cores)
need customized algorithms for computing the degree. The index structures aim
to compactly organize all the cores by decomposing them, or provide insights for
speeding up the query processing. In addition, there is a trade-off between efficiency
and cohesiveness for the first three models. That is, the basic $(k, \mathcal{P})$-core takes the
least running time but may suffer from the issue of low cohesiveness, and the vertex-

Table 5.4 Core-based models and solutions on other general HINs ("—": not applied)

Model	HIN model	Model properties		Vertex type(s)	Algorithm ideas		
		Parameters	Degree		Online	Index-based
Basic $(k, \mathcal{P})$-core [62]	General	k: min degree $\mathcal{P}$: meta-path	# of instances of $\mathcal{P}$	Single	Peeling	Decomposition, tree index
Edge-disjoint $(k, \mathcal{P})$-core [62]	General	k: min degree $\mathcal{P}$: meta-path	# of edge-disjoint instances of $\mathcal{P}$	Single	Peeling max-flow	Decomposition, tree index
Vertex-disjoint $(k, \mathcal{P})$-core [62]	General	k: min degree $\mathcal{P}$: meta-path	# of vertex-disjoint instances of $\mathcal{P}$	Single	Peeling max-flow	Decomposition, tree index
r-com [92]	General	Relational constraints	# of neighbors with a type	Multiple	Peeling	Use round numbers
h-structure [172]	General	—	s-degree (weight all incident edges)	Multiple	h-index	—
$(a_1, \cdots, a_k)$-core [195]	k-partite	$a_i (1 \leq i \leq k)$: min degree	# of neighbors with a type	Multiple	Peeling	—
Multi-layer core [65, 203]	Multi-layer	k (or **k**): min degree(s)	# of neighbors in each layer	Single	Peeling	Decomposition, core cube

Note: k is an integer, and **k** is a vector of integers

Table 5.5 Truss-based models and solutions on other general HINs

Model	HIN model	Model properties			Algorithm
		Parameters	Support	Vertex type(s)	ideas
$(k, \mathcal{P})$-Btruss [180]	General	k: min support $\mathcal{P}$: meta-path	#. of b-triangles w.r.t. $\mathcal{P}$	Single	Peeling, pruning
$(k, \mathcal{P})$-Ctruss [180]	General	k: min support $\mathcal{P}$: meta-path	#. of c-triangles w.r.t. $\mathcal{P}$	Single	Peeling, pruning
$(b_1, \cdots, b_k)$-truss [195]	k-partite	$b_i (1 \leq i \leq k)$: min support	#. of 3-partite-triangles	Multiple	Peeling

disjoint $(k, \mathcal{P})$-core has the highest cohesiveness but takes larger computational cost than others.

5.2.3   Comparison of Different Truss-Based Models

We compare these three HIN truss models in Table 5.5: (1) The $(k, \mathcal{P})$-Btruss and $(k, \mathcal{P})$-Ctruss focus on general HINs, while the $(b_1, \cdots, b_k)$-truss is designed for the multi-partite network. (2) The $(k, \mathcal{P})$-Btruss and $(k, \mathcal{P})$-Ctruss share the same parameters and their vertices are of a single type, but their supports are defined using different triangles (i.e., b- and c-triangles), where each b- or c-triangle has three vertices with the same type. The vertices of the $(b_1, \cdots, b_k)$-truss are with multiple types since each 3-partite-triangle consists of three vertices with three different vertex types. (3) The $(k, \mathcal{P})$-Ctruss is nested within the $(k, \mathcal{P})$-Btruss, since a c-triangle is also a b-triangle but a b-triangle may not be a c-triangle. (4) All the online computation algorithms follow the peeling paradigm, but some customized pruning techniques may be applied. In addition, there is a trade-off between efficiency and cohesiveness for the first two models [180], i.e., the $(k, \mathcal{P})$-Ctruss model tends to exhibit higher cohesiveness, but it is more computationally expensive.

5.2.4   Comparison of Different Clique-Based Models

As presented in Table 5.6, the clique-based models were designed on both the general HIN and its two special cases. Their parameters and link constraints are quite different because they were designed for different application scenarios. The vertices in the cohesive subgraphs are with multiple different types, except for the one on the multi-layer network which only has a single vertex type. Besides, as these models have diverse parameters and link constraints, the algorithms for listing

Table 5.6 Clique-based models and solutions on other general HINs ("—": not applied)

Model	HIN model	Model properties			Algorithm ideas
		Parameters	Link constraints	Vertex type(s)	
Maximal motif clique [81, 101]	General	A motif	Subgraph isomorphism, type-matched vertex set	Multiple	Follow BK algorithm
ABCOutlier-clique [76]	General	A conjunctive select query	Any two vertices with matched types are linked	Multiple	Decompose queries
k-partite clique [47] [74, 115, 129, 139, 195]	k-partite	—	It is a bi-clique between any two linked parties	Multiple	Follow BK algorithm
Multi-layer quasi-clique [19, 20, 138, 183]	Multi-layer	$\gamma_i (i \in [1, l])$: a threshold	It is a quasi-clique in each layer	Single	Use a set enumeration tree

these clique-based subgraphs are different, but two of them follow the idea of BK algorithm which is a classic maximal clique enumeration algorithm on the unipartite network.

Chapter 6
Related Work on CSMs and Solutions

6.1 CSS on Homogeneous Networks

6.1.1 Core-Based CSS

In the literature, the k-core model [145] has been extensively studied and used in various real applications, such as community search [58, 86], recommendation [48], graph mining [63, 124], etc. Many efficient algorithms are developed for computing the k-core in large graphs under various different settings, such as in-memory algorithm [14], disk-based algorithms [38, 95, 175], parallel algorithm [132], and dynamic maintenance algorithms [108, 142, 193]. Besides, the k-core model has been extended for the directed graph, and the extended core models are called Dcore [33, 61, 69, 174] and $[x, y]$-core [124].

In addition, to find the cohesive subgraphs on attributed graphs, the k-core model is also widely used to model the link structure cohesiveness of the subgraph in various attributed graphs, such as keyword-based graphs [54, 56, 57, 186, 194], location-based graphs [55, 60, 166], influence value-based graphs [18, 35, 105, 106], profile-based graphs [37], and so on.

6.1.2 Truss-Based CSS

The k-truss model was independently proposed as the k-dense [141], k-truss [41], triangle k-core [188], and k-community [163]. The subtle difference of these models is the connectedness issue, i.e., the k-dense and triangle k-core models allow the subgraph to be disconnected whereas the k-truss and k-community must be connected subgraphs. Efficient k-truss computation algorithms have been developed under different settings, including in-memory algorithm [164], disk-

© The Author(s), under exclusive license to Springer Nature Switzerland AG 2022
Y. Fang et al., *Cohesive Subgraph Search Over Large Heterogeneous Information Networks*, SpringerBriefs in Computer Science,
https://doi.org/10.1007/978-3-030-97568-5_6

based algorithm [164], distributed algorithm [34], dynamic maintenance algorithms [192, 197]. Besides, the k-truss model has also been extended for the directed graph, and the extended model is called D-truss [116].

Similar to the k-core model above, the k-truss model has also been used for modeling the link structure cohesiveness in searching cohesive subgraphs over attributed graphs, such as simple graphs [84, 87], keyword-based graphs [85, 205], location-based graphs [32], and general attributed graphs [117].

6.1.3   Clique-Based CSS

The k-clique has been widely used in various areas, such as communication network analysis, chemical data analysis, protein structure prediction, and so on. To list the k-cliques in a large graph for a specific integer k, many efficient enumeration algorithms [42, 104] have been developed. Another group of works is to study efficient algorithms for enumerating maximal cliques in large graphs, such as in-memory algorithms [161], external-memory algorithms [39], and parallel algorithms [144], and a detailed survey can be found in [22]. Meanwhile, as the condition of k-clique is strict, some relaxed variants [58] such as k-plex and quasi-clique are proposed to identify cohesive subgraphs. For example, Devora et al. [17] studied how to efficiently enumerate maximal k-plexes in large graphs, and Zeng et al. developed algorithms for discovering coherent closed quasi-clique [184] from large dense graph databases. In addition, the clique is also used for finding cohesive subgraphs from attributed graphs [58, 186].

6.1.4   Connectivity-Based CSS

This kind of subgraphs [70] has attracted much research attention and used in some real applications [26] such as social behavior mining, web data analysis, etc. In [178], Yan et al. investigated the problem of computing frequent closed k-ECCs from a set of data graphs based on the k-edge connectivity. In [26, 196] and [25], both efficient online algorithms and index-based algorithms are presented for computing the k-ECCs in large graphs. In [82, 83], hu et al. investigated the problem of finding the minimal Steiner maximum-connected subgraph (SMCS), where the SMCS is the subgraph with the largest connectivity in a graph.

6.1.5   Density-Based CSS

The densest subgraph discovery (DSD) is a fundamental problem in graph mining, and the densest subgraph (CS) is widely used in network science, biological analysis, graph databases, and system optimization. The DSD problem can be addressed exactly by solving a parametric maximum-flow problem [67, 73]. Nevertheless, the exact DSD algorithms are costly since they have high time complexities, so some efficient approximation DSD algorithms are developed. In [27, 63], 0.5-approximation algorithms are developed which only take linear time cost. Recently, the density-friendly graph decomposition problem [160] has been studied, which decomposes a graph into a sequence of subgraphs in terms of their density values [43, 160].

Additionally, many variants of the DS discovery problem have been investigated. In [140], Qin et al. proposed to find the top-k locally densest subgraphs. In [10], Bahmani et al. devised an efficient approximation algorithm to solve the DS discovery under the streaming model. In [63, 131, 162], the k-clique DS problem is studied, which extends definition of edge-density (see Definition 2.9) by considering a k-clique. In [10, 27, 124–126], the DS problem on directed graphs is studied.

6.2 HIN Clustering

In this section, we review another group of highly related work to our survey, called HIN clustering or HIN community detection (CD), which generally aims to cluster all the vertices of the network into some sets such that each set of vertices is densely connected internally. *Although this group of works also generate cohesive subgraphs in the HIN, it is different from the works in our survey and the three key differences, namely problem settings, criteria of defining cohesive subgraphs, and algorithms, have been highlighted in Sect. 1.*

6.2.1 Bipartite Networks

In the literature, there are many works studying community detection in bipartite networks [12, 16, 50, 75, 99, 119, 201, 202]. For lack of space, here we mainly review two representative groups of such works. The first group detects communities by extending the classic unipartite modularity for bipartite networks. Barber et al. [12] introduced the concept of bipartite modularity, and proposed an algorithm BRIM for identifying network modules. Guimerà [75] also introduced a new modularity model that can be applied to identify modules in both bipartite networks and directed unipartite networks. Liu et al. [119] developed another community detection algorithm LP&BRIM, which is based on a joint strategy of label propagation (LP) and BRIM. Focused on weighted bipartite networks, QuanBiMo [50] detects communities by maximizing the weighted modularity. Later on, Beckett et al. [16] introduced two new algorithms, LPAwb+ and DIRTLPAwb+, for maximizing weighted modularity in bipartite networks. The second group of works is based on the stochastic block model (SBM). In [99], Larremore et al. formulated a bipartite SBM, yielding a projection-free and statistically principled method for community detection. Zhou et al. studied the spectral clustering algorithms [201] for community detection under a general bipartite SBM. Later, they presented an optimal two-stage procedure for bipartite network clustering and established its weak consistency under a general bipartite SBM [202].

6.2.2   Other General HINs

In recent years, some works [147, 148, 152, 154, 155, 157, 199] have focused on generating clusters/communities in the HIN, which can be classified into two groups according to vertex types in the communities. The first group [31, 148, 152, 157] focuses on detecting clusters, each of which contains objects with multiple types. In [157], Sun et al. clustered star-schema HINs by assuming target objects to be generated by a ranking-based probabilistic generative model. In [152], Sun et al. designed a probabilistic clustering model which considers the strengths of different link types. In [148], Shi et al. projected the HIN into many sub-networks and estimated the reachable probability of objects in each sub-network for clustering. In [31], Chen et al. studied the clustering of star-schema HIN. The second group [120, 154, 155, 199] aims to generate clusters of objects with a specific type. In [154], Sun et al. proposed an algorithm to generate clusters of objects with a specific type; in [155], a user-guided algorithm is developed to cluster objects of a target type; in [199], a social influence-based algorithm is presented to cluster objects of a specific type in the graph; in [120], a graph neural network-based model is developed for detecting communities of objects with a target type in the HIN.

6.2.3   Comparison with the Earlier Version

An earlier version of this book has been accepted as an ACM SIGMOD'2021 tutorial [59]. Compared to the old version, this book has several newly added contributions, summarized as follows:

1. We show more background information of CSS over large HINs, such as typical examples of HINs, applications of CSS, references, and so on.
2. We classify the CSMs on HINs into different categories and systemically review the solutions for the models in each category especially the state-of-the-art algorithms.
3. We extensively analyze and compare existing CSMs and their corresponding solutions over HINs from various angles.
4. We further review the existing works in the literature that are highly related to our survey including CSS over homogeneous networks and HIN clustering.

Chapter 7
Future Work and Conclusion

7.1 Novel Application-Driven CSMs

As shown in Chaps. 3 and 4, there are several CSMs on HINs, but they are still insufficient to cover all related real applications, especially for general HINs that contain various types of vertices and edges. This is because most of these models are extended from classic CSMs on homogeneous networks. However, for domain-specific HINs such as medical knowledge graphs, the applications of cohesive subgraphs are often different with those on homogeneous networks. For example, in medical knowledge graphs [28], the cohesive subgraphs may reveal the groups of users and animals that are infected with COVID-19, but these subgraphs may not be captured by existing CSMs. Thus, it is challenging to develop novel cohesive models, by carefully considering the need in real applications on HINs.

To develop a novel application-driven CSM, the first important research direction is to develop an application-based evaluation metric, such that it can be used to evaluate the quality of the subgraph of the CSM in terms of the specific application. Notice that currently, there is no standard metric that can be used to evaluate the quality of different CSMs. Based on the application-based metric, novel models could be developed to maximize the metric values in the applications. During this process, existing CSM models could provide some insights for formulating novel models. Finally, the proposed application-driven model should be further tuned by evaluating its performance on the applications in real-world HINs.

7.2 Efficient Search Algorithms on Big HINs

As aforementioned, a key challenge of CSS over big HINs is how to improve the efficiency. Although existing works have made plenty of effort to alleviate the

© The Author(s), under exclusive license to Springer Nature Switzerland AG 2022   61
Y. Fang et al., *Cohesive Subgraph Search Over Large Heterogeneous Information Networks*, SpringerBriefs in Computer Science,
https://doi.org/10.1007/978-3-030-97568-5_7

issues of efficiency and scalability, there is much room for further improving the search algorithm so that we can better tame the sheer volume and complex structure in searching large scale HINs. To develop efficient algorithms, there are several potential research directions:

1. *Parallel algorithms.* Parallel algorithms often exploit a distributed computing platform or multi-core computing resources to speedup the computation. They have been widely developed for searching cohesive subgraphs on homogeneous networks. For example, to enable efficient parallel k-core computation, many parallel algorithms (e.g., [24, 46, 53, 90, 93, 132, 176]) have been developed. Consequently, it is a promising research direction to improve the efficiency.

2. *Approximation algorithms.* For many CSS problems (e.g., densest subgraph discovery), the exact algorithms may have high time complexities, rendering them impractical for large-scale networks. To alleviate this issue, researchers often resort to approximation algorithms by trading the accuracy. For instance, to find the densest subgraph, the exact algorithms (e.g., undirected graphs [63, 73, 131, 162] and directed graphs [27, 96, 124]) need to compute the maximum flow of a network which is very slow, while the approximation algorithms could be much simpler and faster (e.g., undirected graphs [10, 27, 63] and directed graphs [10, 27, 94, 96, 124]). Hence, it is an important means of reducing the time cost.

3. *I/O-efficient algorithms.* When the network datasets cannot be fitted into the memory of a single machine, the I/O-efficient query algorithms (e.g., [107]) can make the subgraph search possible. In other words, developing I/O-efficient algorithms may provide us an option to tame the sheer volume of HINs.

7.3   Parameters Optimization

Almost all the existing solutions of CSS require users to input a list of query parameters (e.g., k and $\mathcal{P}$ in Table 1.1), and also often assume that query users can manually specify proper values for these parameters in the queries. This assumption, however, may not make much sense, especially when they are not familiar with the network. For example, in the $(k, \mathcal{P})$-core model [62], if k is set too large, then there may not exist a corresponding $(k, \mathcal{P})$-core in the HIN; if k is set too small, the $(k, \mathcal{P})$-core may not be of much interest, since it is very loosely connected. Consequently, an important issue of CSS on HINs is to automatically optimize the query parameters so that some proper parameter values can be suggested. Recently, some attempts have been made to address this issue. For example, Chu et al. [40] have studied how to select the best k in core computation, which is able to suggest the best value of k in terms of a specific scoring metric. However, this work mainly focuses on classic homogeneous networks, and it is not clear how to adapt this method for processing HINs. Meanwhile, to the best of our knowledge, none of the existing works has tried to tackle this issue on HINs. Hence, it is desirable to study how to optimize the query parameters for CSS over HINs.

7.4 An Online Repository for Collecting HIN Datasets, Tools, and Algorithm Codes

There are several online platforms for keeping large network datasets, such as SNAP,[1] LAW,[2] and KONECT,[3] which maintain a large number of network datasets and allow online users to download them easily. However, most of these networks are homogeneous networks and only a few HIN datasets are collected into these platforms. On the other hand, the research on HINs has gained increasing attention in recent years. Hence, it is necessary to build a dedicated online repository for keeping and maintaining various real HIN datasets, which will make researchers and practitioners easily access these datasets, but also provide a benchmark for academic research. In addition, other HIN related resources, such as visualization tools and algorithms codes, could be also collected in this repository.

7.5 Conclusion

In this book, we conduct a comprehensive survey on the topic of CSS over large HINs. We systematically review over 40 research articles, which focus on this topic, published between 1996 and 2021. More specifically, we first classify existing studies according to the network types and CSMs. Then, we review and discuss the representative studies on each type of HINs. Furthermore, we analyze and compare these works in terms of computational cost and cohesiveness. Finally, we point out a list of future research directions. In summary, our survey provides an overview of the state-of-the-art research on the topic of CSS over large HINs, which will give researchers a thorough understanding of this topic.

---

[1] https://snap.stanford.edu/data/.

[2] http://law.di.unimi.it/.

[3] http://konect.cc/.

References

1. Abello, J., Resende, M. G., & Sudarsky, S. (2002). Massive quasi-clique detection. In *Latin American symposium on theoretical informatics* (pp. 598–612). Springer.
2. Abidi, A., Chen, L., Zhou, R., & Liu, C. (2021). Searching personalized k-wing in large and dynamic bipartite graphs. arXiv preprint arXiv:2101.00810.
3. Abidi, A., Zhou, R., Chen, L., & Liu, C. (2020). Pivot-based maximal biclique enumeration. In *IJCAI* (pp. 3558–3564).
4. Acuña, V., Ferreira, C. E., Freire, A. S., & Moreno, E. (2014). Solving the maximum edge biclique packing problem on unbalanced bipartite graphs. *Discrete Applied Mathematics, 164*, 2–12.
5. Ahmed, A., Batagelj, V., Fu, X., Hong, S.-H., Merrick, D., & Mrvar, A. (2007). Visualisation and analysis of the internet movie database. In *2007 6th International Asia-Pacific Symposium on Visualization* (pp. 17–24). IEEE.
6. Ahuja, R. K., Orlin, J. B., Stein, C., & Tarjan, R. E. (1994). Improved algorithms for bipartite network flow. *SIAM Journal on Computing, 23*(5), 906–933.
7. Akoglu, L., Tong, H., & Koutra, D. (2015). Graph based anomaly detection and description: A survey. *Data Mining and Knowledge Discovery, 29*(3), 626–688.
8. Al-Yamani, A. A., Ramsundar, S., & Pradhan, D. K. (2007). A defect tolerance scheme for nanotechnology circuits. *IEEE Transactions on Circuits and Systems I: Regular Papers, 54*(11), 2402–2409.
9. Andersen, R. (2010). A local algorithm for finding dense subgraphs. *ACM Transactions on Algorithms, 6*(4), 1–12.
10. Bahmani, B., Kumar, R., & Vassilvitskii, S. (2012). Densest subgraph in streaming and mapreduce. *PVLDB, 5*(5), 454–465.
11. Ban, Y., & Duan, Y. (2018). On finding dense subgraphs in bipartite graphs: Linear algorithms. arXiv preprint arXiv:1810.06809.
12. Barber, M. J. (2007). Modularity and community detection in bipartite networks. *Physical Review E, 76*(6), 066102.
13. Barman, D., Bhattacharya, S., Sarkar, R., & Chowdhury, N. (2019). k-context technique: A method for identifying dense subgraphs in a heterogeneous information network. *IEEE Transactions on Computational Social Systems, 6*(6), 1190–1205.
14. Batagelj, V., & Zaversnik, M. (2003). An o (m) algorithm for cores decomposition of networks. arXiv preprint cs/0310049.
15. Batagelj, V., & Zaveršnik, M. (2011). Fast algorithms for determining (generalized) core groups in social networks. *Advances in Data Analysis and Classification, 5*(2), 129–145.

© The Author(s), under exclusive license to Springer Nature Switzerland AG 2022   65
Y. Fang et al., *Cohesive Subgraph Search Over Large Heterogeneous
Information Networks*, SpringerBriefs in Computer Science,
https://doi.org/10.1007/978-3-030-97568-5

16. Beckett, S. J. (2016). Improved community detection in weighted bipartite networks. *Royal Society Open Science, 3*(1), 140536.
17. Berlowitz, D., Cohen, S., & Kimelfeld, B. (2015). Efficient enumeration of maximal k-plexes. In *SIGMOD* (pp. 431–444). ACM.
18. Bi, F., Chang, L., Lin, X., & Zhang, W. (2018). An optimal and progressive approach to online search of top-k influential communities. *PVLDB, 11*(9), 1056–1068.
19. Boden, B., Günnemann, S., Hoffmann, H., & Seidl, T. (2012). Mining coherent subgraphs in multi-layer graphs with edge labels. In *SIGKDD* (pp. 1258–1266). ACM.
20. Boden, B., Günnemann, S., Hoffmann, H., & Seidl, T. (2017). Mimag: Mining coherent subgraphs in multi-layer graphs with edge labels. *Knowledge and Information Systems, 50*(2), 417–446.
21. Bron, C., & Kerbosch, J. (1973). Algorithm 457: Finding all cliques of an undirected graph. *Communications of the ACM, 16*(9), 575–576.
22. Cazals, F., & Karande, C. (2008). A note on the problem of reporting maximal cliques. *Theoretical Computer Science, 407*(1–3), 564–568.
23. Cerinšek, M., & Batagelj, V. (2015). Generalized two-mode cores. *Social Networks, 42*, 80–87.
24. Chan, T. H., Sozio, M., & Sun, B. (2019). Distributed approximate k-core decomposition and min-max edge orientation: Breaking the diameter barrier. In *IEEE International Parallel and Distributed Processing Symposium (IPDPS)* (pp. 345–354). IEEE.
25. Chang, L., Lin, X., Qin, L., Yu, J. X., & Zhang, W. (2015). Index-based optimal algorithms for computing Steiner components with maximum connectivity. In *SIGMOD* (pp. 459–474). ACM.
26. Chang, L., Yu, J. X., Qin, L., Lin, X., Liu, C., & Liang, W. (2013). Efficiently computing k-edge connected components via graph decomposition. In *SIGMOD* (pp. 205–216). ACM.
27. Charikar, M. (2000). Greedy approximation algorithms for finding dense components in a graph. In *International Workshop on Approximation Algorithms for Combinatorial Optimization* (pp. 84–95). Springer.
28. Chatterjee, A., Nardi, C., Oberije, C., & Lambin, P. (2021). Knowledge graphs for covid-19: An exploratory review of the current landscape. *Journal of Personalized Medicine, 11*(4), 300.
29. Chen, C., Zhu, Q., Wu, Y., Sun, R., Wang, X., & Liu, X. (2021). Efficient critical relationships identification in bipartite networks. In *World Wide Web* (pp. 1–21).
30. Chen, H., & Liu, T. (2017). Maximum edge bicliques in tree convex bipartite graphs. In *International Workshop on Frontiers in Algorithmics* (pp. 47–55). Springer.
31. Chen, L., Gao, Y., Zhang, Y., Jensen, C. S., & Zheng, B. (2019). Efficient and incremental clustering algorithms on star-schema heterogeneous graphs. In *ICDE* (pp. 256–267). IEEE.
32. Chen, L., Liu, C., Zhou, R., Li, J., Yang, X., & Wang, B. (2018). Maximum co-located community search in large scale social networks. *PVLDB, 11*(10), 1233–1246.
33. Chen, L., Liu, C., Zhou, R., Xu, J., & Li, J. (2021). Efficient exact algorithms for maximum balanced biclique search in bipartite graphs (pp. 248–260).
34. Chen, P.-L., Chou, C.-K., & Chen, M.-S. (2014). Distributed algorithms for k-truss decomposition. In *Big data* (pp. 471–480). IEEE.
35. Chen, S., Wei, R., Popova, D., & Thomo, A. (2016). Efficient computation of importance based communities in web-scale networks using a single machine. In *CIKM* (pp. 1553–1562).
36. Chen, X., Wang, K., Lin, X., Zhang, W., Qin, L., & Zhang, Y. (2021). Efficiently answering reachability and path queries on temporal bipartite graphs. *Proceedings of the VLDB Endowment, 14*(10), 1845–1858.
37. Chen, Y., Fang, Y., Cheng, R., Li, Y., Chen, X., & Zhang, J. (2018). Exploring communities in large profiled graphs. *IEEE Transactions on Knowledge and Data Engineering, 31*(8), 1624–1629.
38. Cheng, J., Ke, Y., Chu, S., & Özsu, M. T. (2011). Efficient core decomposition in massive networks. In *ICDE* (pp. 51–62). IEEE.
39. Cheng, J., Ke, Y., Fu, A. W.-C., Yu, J. X., & Zhu, L. (2011). Finding maximal cliques in massive networks. *TODS, 36*(4), 1–34.

40. Chu, D., Zhang, F., Lin, X., Zhang, W., Zhang, Y., Xia, Y., & Zhang, C. (2020). Finding the best k in core decomposition: A time and space optimal solution. In *ICDE* (pp. 685–696). IEEE.
41. Cohen, J. (2008). Trusses: Cohesive subgraphs for social network analysis. *National security Agency Technical Report, 16*, 3–1.
42. Danisch, M., Balalau, O., & Sozio, M. (2018). Listing k-cliques in sparse real-world graphs. In *WWW* (pp. 589–598).
43. Danisch, M., Chan, T.-H. H., & Sozio, M. (2017). Large scale density-friendly graph decomposition via convex programming. In *WWW* (pp. 233–242).
44. Das, A., & Tirthapura, S. (2018). Incremental maintenance of maximal bicliques in a dynamic bipartite graph. *IEEE Transactions on Multi-Scale Computing Systems, 4*(3), 231–242.
45. Das, A., & Tirthapura, S. (2019). Shared-memory parallel maximal biclique enumeration. In *HiPC* (pp. 34–43).
46. Dasari, N. S., Desh, R., & Zubair, M. (2014). Park: An efficient algorithm for k-core decomposition on multicore processors. In *IEEE International Conference on Big Data (Big Data)* (pp. 9–16). IEEE.
47. Dawande, M., Keskinocak, P., Swaminathan, J. M., & Tayur, S. (2001). On bipartite and multipartite clique problems. *Journal of Algorithms, 41*(2), 388–403.
48. Ding, D., Li, H., Huang, Z., & Mamoulis, N. (2017). Efficient fault-tolerant group recommendation using alpha-beta-core. In *CIKM* (pp. 2047–2050).
49. Ding, Q., Katenka, N., Barford, P., Kolaczyk, E., & Crovella, M. (2012). Intrusion as (anti) social communication: Characterization and detection. In *SIGKDD* (pp. 886–894). ACM.
50. Dormann, C. F., & Strauss, R. (2014). A method for detecting modules in quantitative bipartite networks. *Methods in Ecology and Evolution, 5*(1), 90–98.
51. Dudley, J. T., Deshpande, T., & Butte, A. J. (2011). Exploiting drug–disease relationships for computational drug repositioning. *Briefings in Bioinformatics, 12*(4), 303–311.
52. Eppstein, D. (1994). Arboricity and bipartite subgraph listing algorithms. *Information Processing Letters, 51*(4), 207–211.
53. Esfandiari, H., Lattanzi, S., & Mirrokni, V. (2018). Parallel and streaming algorithms for k-core decomposition. arXiv preprint arXiv:1808.02546.
54. Fang, Y., Cheng, R., Chen, Y., Luo, S., & Hu, J. (2017). Effective and efficient attributed community search. *The VLDB Journal, 26*(6), 803–828.
55. Fang, Y., Cheng, R., Li, X., Luo, S., & Hu, J. (2017). Effective community search over large spatial graphs. *PVLDB, 10*(6), 709–720.
56. Fang, Y., Cheng, R., Luo, S., & Hu, J. (2016). Effective community search for large attributed graphs. *PVLDB, 9*(12), 1233–1244.
57. Fang, Y., Cheng, R., Luo, S., Hu, J., & Huang, K. (2017). C-explorer: Browsing communities in large graphs. *PVLDB, 10*(12), 1885–1888.
58. Fang, Y., Huang, X., Qin, L., Zhang, Y., Zhang, W., Cheng, R., & Lin, X. (2020). A survey of community search over big graphs. *The VLDB Journal, 29*(1), 353–392.
59. Fang, Y., Wang, K., Lin, X., & Zhang, W. (2021). Cohesive subgraph search over big heterogeneous information networks: Applications, challenges, and solutions. *ACM SIGMOD* (pp. 2829–2838).
60. Fang, Y., Wang, Z., Cheng, R., Li, X., Luo, S., Hu, J., & Chen, X. (2019). On spatial-aware community search. *TKDE, 31*(4), 783–798.
61. Fang, Y., Wang, Z., Cheng, R., Wang, H., & Hu, J. (2019). Effective and efficient community search over large directed graphs. *TKDE, 31*(11), 2093–2107.
62. Fang, Y., Yang, Y., Zhang, W., Lin, X., & Cao, X. (2020). Effective and efficient community search over large heterogeneous information networks. *PVLDB, 13*(6), 854–867.
63. Fang, Y., Yu, K., Cheng, R., Lakshmanan, L. V., & Lin, X. (2019). Efficient algorithms for densest subgraph discovery. *PVLDB, 12*(11), 1719–1732.
64. Fang, Y., Zhang, H., Ye, Y., & Li, X. (2014). Detecting hot topics from twitter: A multiview approach. *Journal of Information Science, 40*(5), 578–593.

65. Galimberti, E., Bonchi, F., & Gullo, F. (2017). Core decomposition and densest subgraph in multilayer networks. In *CIKM* (pp. 1807–1816).
66. Galimberti, E., Bonchi, F., Gullo, F., & Lanciano, T. (2020). Core decomposition in multilayer networks: Theory, algorithms, and applications. *ACM TKDD, 14*(1), 1–40.
67. Gallo, G., Grigoriadis, M. D., & Tarjan, R. E. (1989). A fast parametric maximum flow algorithm and applications. *SIAM Journal on Computing, 18*(1), 30–55.
68. Giatsidis, C., Thilikos, D. M., & Vazirgiannis, M. (2011). Evaluating cooperation in communities with the k-core structure. In *ASONAM* (pp. 87–93). IEEE.
69. Giatsidis, C., Thilikos, D. M., & Vazirgiannis, M. (2013). D-cores: Measuring collaboration of directed graphs based on degeneracy. *Knowledge and Information Systems, 35*(2), 311–343.
70. Gibbons, A. (1985). *Algorithmic graph theory*. Cambridge University Press.
71. Gionis, A., & Tsourakakis, C. E. (2015). Dense subgraph discovery: Kdd 2015 tutorial. In *SIGKDD* (pp. 2313–2314). New York, NY: Association for Computing Machinery.
72. Glover, F. (1997). Tabu search and adaptive memory programmin—advances, applications and challenges. In *Interfaces in computer science and operations research* (pp. 1–75). Springer.
73. Goldberg, A. V. (1984). *Finding a maximum density subgraph*. University of California Berkeley, CA.
74. Grünert, T., Irnich, S., Zimmermann, H.-J., Schneider, M., & Wulfhorst, B. (2002). Finding all k-cliques in k-partite graphs, an application in textile engineering. *Computers & Operations Research, 29*(1), 13–31.
75. Guimerà, R., Sales-Pardo, M., & Amaral, L. A. N. (2007). Module identification in bipartite and directed networks. *Physical Review E, 76*(3), 036102.
76. Gupta, M., Gao, J., Yan, X., Cam, H., & Han, J. (2013). On detecting association-based clique outliers in heterogeneous information networks. In *ASONAM* (pp. 108–115). IEEE.
77. Hao, Y., Zhang, M., Wang, X., & Chen, C. (2020). Cohesive subgraph detection in large bipartite networks. In *International Conference on Scientific and Statistical Database Management* (pp. 1–4).
78. Hartmanis, J. (1982). Computers and intractability: A guide to the theory of np-completeness (Michael R. Garey and David S. Johnson). *Siam Review, 24*(1), 90.
79. He, Y., Wang, K., Zhang, W., Lin, X., & Zhang, Y. (2021). Exploring cohesive subgraphs with vertex engagement and tie strength in bipartite graphs. *Information Sciences, 572*, 277–296.
80. Hirsch, J. E. (2005). An index to quantify an individual's scientific research output. *Proceedings of the National academy of Sciences, 102*(46), 16569–16572.
81. Hu, J., Cheng, R., Chang, K. C.-C., Sankar, A., Fang, Y., & Lam, B. Y. (2019). Discovering maximal motif cliques in large heterogeneous information networks. In *ICDE* (pp. 746–757). IEEE.
82. Hu, J., Wu, X., Cheng, R., Luo, S., & Fang, Y. (2016). Querying minimal Steiner maximum-connected subgraphs in large graphs. In *CIKM* (pp. 1241–1250).
83. Hu, J., Wu, X., Cheng, R., Luo, S., & Fang, Y. (2017). On minimal Steiner maximum-connected subgraph queries. *TKDE, 29*(11), 2455–2469.
84. Huang, X., Cheng, H., Qin, L., Tian, W., & Yu, J. X. (2014). Querying k-truss community in large and dynamic graphs. In *SIGMOD* (pp. 1311–1322). ACM.
85. Huang, X., & Lakshmanan, L. V. (2017). Attribute-driven community search. *PVLDB, 10*(9), 949–960.
86. Huang, X., Lakshmanan, L. V., & Xu, J. (2017). Community search over big graphs: Models, algorithms, and opportunities. In *ICDE* (pp. 1451–1454). IEEE.
87. Huang, X., Lakshmanan, L. V., Yu, J. X., & Cheng, H. (2015). Approximate closest community search in networks. *PVLDB, 9*(4).
88. Huang, Z., Zheng, Y., Cheng, R., Sun, Y., Mamoulis, N., & Li, X. (2016). Meta structure: Computing relevance in large heterogeneous information networks. In *SIGKDD* (pp. 1595–1604). ACM.

89. Ignatov, D. I., Ivanova, P., & Zamaletdinova, A. (2018). Mixed integer programming for searching maximum quasi-bicliques. In *International Conference on Network Analysis* (pp. 19–35). Springer.
90. Jakma, P., Orczyk, M., Perkins, C. S., & Fayed, M. (2012). Distributed k-core decomposition of dynamic graphs. In *Proceedings of the 2012 ACM Conference on CoNEXT Student Workshop* (pp. 39–40).
91. Jethava, V., & Beerenwinkel, N. (2015). Finding dense subgraphs in relational graphs. In *Joint European Conference on Machine Learning and Knowledge Discovery in Databases* (pp. 641–654). Springer.
92. Jian, X., Wang, Y., & Chen, L. (2020). Effective and efficient relational community detection and search in large dynamic heterogeneous information networks. *PVLDB, 13*(10), 1723–1736.
93. Kabir, H., & Madduri, K. (2017). Parallel k-core decomposition on multicore platforms. In *IEEE International Parallel and Distributed Processing Symposium Workshops (IPDPSW)* (pp. 1482–1491). IEEE.
94. Kannan, R., & Vinay, V. (1999). *Analyzing the structure of large graphs.* Rheinische Friedrich-Wilhelms-Universität Bonn, Bonn.
95. Khaouid, W., Barsky, M., Srinivasan, V., & Thomo, A. (2015). K-core decomposition of large networks on a single pc. *PVLDB, 9*(1), 13–23.
96. Khuller, S., & Saha, B. (2009). On finding dense subgraphs. In *International Colloquium on Automata, Languages, and Programming* (pp. 597–608). Springer.
97. Kumar, R., Tomkins, A., & Vee, E. (2008). Connectivity structure of bipartite graphs via the knc-plot. In *WSDM* (pp. 129–138).
98. Lakhotia, K., Kannan, R., Prasanna, V., & De Rose, C. A. (2021). Receipt: Refine coarse-grained independent tasks for parallel tip decomposition of bipartite graphs. *PVLDB, 14*(3), 404–417.
99. Larremore, D. B., Clauset, A., & Jacobs, A. Z. (2014). Efficiently inferring community structure in bipartite networks. *Physical Review E, 90*(1), 012805.
100. Ley, M. (2002). The DBLP computer science bibliography: Evolution, research issues, perspectives. In *String Processing and Information Retrieval, 9th International Symposium, SPIRE 2002, Lisbon, Portugal, September 11–13, 2002, Proceedings* (pp. 1–10).
101. Li, B., Cheng, R., Hu, J., Fang, Y., Ou, M., Luo, R., Chang, K. C.-C., & Lin, X. (2020). Mc-explorer: Analyzing and visualizing motif-cliques on large networks. In *ICDE* (pp. 1–12). IEEE.
102. Li, J., Liu, G., Li, H., & Wong, L. (2007). Maximal biclique subgraphs and closed pattern pairs of the adjacency matrix: A one-to-one correspondence and mining algorithms. *TKDE, 19*(12), 1625–1637.
103. Li, M., Hao, J.-K., & Wu, Q. (2020). General swap-based multiple neighborhood adaptive search for the maximum balanced biclique problem. *Computers & Operations Research, 119,* 104922.
104. Li, R.-H., Gao, S., Qin, L., Wang, G., Yang, W., & Yu, J. X. (2020). Ordering heuristics for k-clique listing. *PVLDB, 13*(12), 2536–2548.
105. Li, R.-H., Qin, L., Ye, F., Yu, J. X., Xiao, X., Xiao, N., & Zheng, Z. (2018). Skyline community search in multi-valued networks. In *SIGMOD* (pp. 457–472).
106. Li, R.-H., Qin, L., Yu, J. X., & Mao, R. (2015). Influential community search in large networks. *PVLDB, 8*(5), 509–520.
107. Li, R.-H., Qin, L., Yu, J. X., & Mao, R. (2017). Finding influential communities in massive networks. *The VLDB Journal, 26*(6), 751–776.
108. Li, R.-H., Yu, J. X., & Mao, R. (2013). Efficient core maintenance in large dynamic graphs. *TKDE, 26*(10), 2453–2465.
109. Li, Y., Kuboyama, T., & Sakamoto, H. (2013). Truss decomposition for extracting communities in bipartite graph. In *Third International Conference on Advances in Information Mining and Management* (pp. 76–80).

110. Linghu, Q., Zhang, F., Lin, X., Zhang, W., & Zhang, Y. (2020). Global reinforcement of social networks: The anchored coreness problem. In *SIGMOD* (pp. 2211–2226).

111. Liu, B., Yuan, L., Lin, X., Qin, L., Zhang, W., & Zhou, J. (2019). Efficient $(\alpha,\ \beta)$-core computation: An index-based approach. In *WWW* (pp. 1130–1141).

112. Liu, B., Yuan, L., Lin, X., Qin, L., Zhang, W., & Zhou, J. (2020). Efficient $(\alpha,\ \beta)$-core computation in bipartite graphs. *The VLDB Journal, 29*(5), 1075–1099.

113. Liu, B., Zhang, F., Zhang, C., Zhang, W., & Lin, X. (2019). Corecube: Core decomposition in multilayer graphs. In *WISE* (pp. 694–710). Springer.

114. Liu, G., Sim, K., & Li, J. (2006). Efficient mining of large maximal bicliques. In *International Conference on Data Warehousing and Knowledge Discovery* (pp. 437–448). Springer.

115. Liu, Q., Chen, Y.-P. P., & Li, J. (2014). k-partite cliques of protein interactions: A novel subgraph topology for functional coherence analysis on ppi networks. *Journal of Theoretical Biology, 340*, 146–154.

116. Liu, Q., Zhao, M., Huang, X., Xu, J., & Gao, Y. (2020). Truss-based community search over large directed graphs. In *SIGMOD* (pp. 2183–2197).

117. Liu, Q., Zhu, Y., Zhao, M., Huang, X., Xu, J., and Gao, Y. (2020). Vac: Vertex-centric attributed community search. In *ICDE* (pp. 937–948). IEEE.

118. Liu, X., Li, J., & Wang, L. (2008). Modeling protein interacting groups by quasi-bicliques: Complexity, algorithm, and application. *IEEE/ACM Transactions on Computational Biology and Bioinformatics, 7*(2), 354–364.

119. Liu, X., & Murata, T. (2010). Community detection in large-scale bipartite networks. *Transactions of the Japanese Society for Artificial Intelligence, 25*(1), 16–24.

120. Luo, L., Fang, Y., Cao, X., Zhang, X., & Zhang, W. (2021). Detecting communities from heterogeneous graphs: A context path-based graph neural network model. In *CIKM* (pp. 1170–1180).

121. Luo, W., Zhou, X., Li, K., Gao, Y., & Li, K. (2021). Efficient influential community search in large uncertain graphs. *TKDE*.

122. Luo, W., Zhou, X., Yang, J., Peng, P., Xiao, G., & Gao, Y. (2020). Efficient approaches to top-r influential community search. *IEEE Internet of Things Journal, 8*(16), 12650–12657.

123. Lyu, B., Qin, L., Lin, X., Zhang, Y., Qian, Z., & Zhou, J. (2020). Maximum biclique search at billion scale. *PVLDB, 13*(9), 1359–1372.

124. Ma, C., Fang, Y., Cheng, R., Lakshmanan, L. V., Zhang, W., & Lin, X. (2020). Efficient algorithms for densest subgraph discovery on large directed graphs. In *SIGMOD* (pp. 1051–1066). ACM.

125. Ma, C., Fang, Y., Cheng, R., Lakshmanan, L. V., Zhang, W., & Lin, X. (2021). Efficient directed densest subgraph discovery. *ACM SIGMOD Record, 50*(1), 33–40.

126. Ma, C., Fang, Y., Cheng, R., Lakshmanan, L. V., Zhang, W., & Lin, X. (2021). On directed densest subgraph discovery. *ACM Transactions on Database Systems (TODS), 46*(4), 1–45.

127. Ma, Z., Liu, Y., Hu, Y., Yang, J., Liu, C., & Dai, H. (2021). Efficient maintenance for maximal bicliques in bipartite graph streams. In *World Wide Web* (pp. 1–21).

128. McCreesh, C., & Prosser, P. (2014). An exact branch and bound algorithm with symmetry breaking for the maximum balanced induced biclique problem. In *International Conference on AI and OR Techniques in Constraint Programming for Combinatorial Optimization Problems* (pp. 226–234). Springer.

129. Mirghorbani, M., & Krokhmal, P. (2013). On finding k-cliques in k-partite graphs. *Optimization Letters, 7*(6), 1155–1165.

130. Mishra, N., Ron, D., & Swaminathan, R. (2004). A new conceptual clustering framework. *Machine Learning, 56*(1–3), 115–151.

131. Mitzenmacher, M., Pachocki, J., Peng, R., Tsourakakis, C., & Xu, S. C. (2015). Scalable large near-clique detection in large-scale networks via sampling. In *SIGKDD* (pp. 815–824). ACM.

132. Montresor, A., De Pellegrini, F., & Miorandi, D. (2012). Distributed k-core decomposition. *IEEE TPDS, 24*(2), 288–300.

133. Mukherjee, A. P., & Tirthapura, S. (2016). Enumerating maximal bicliques from a large graph using mapreduce. *IEEE Transactions on Services Computing, 10*(5), 771–784.

134. Nussbaum, D., Pu, S., Sack, J.-R., Uno, T., & Zarrabi-Zadeh, H. (2012). Finding maximum edge bicliques in convex bipartite graphs. *Algorithmica, 64*(2), 311–325.

135. Pandey, A., Sharma, G., & Jain, N. (2020). Maximum weighted edge biclique problem on bipartite graphs. In *Conference on Algorithms and Discrete Applied Mathematics* (pp. 116–128). Springer.

136. Pavlopoulos, G. A., Kontou, P. I., Pavlopoulou, A., Bouyioukos, C., Markou, E., & Bagos, P. G. (2018). Bipartite graphs in systems biology and medicine: A survey of methods and applications. *GigaScience, 7*(4), giy014.

137. Peeters, R. (2003). The maximum edge biclique problem is np-complete. *Discrete Applied Mathematics, 131*(3), 651–654.

138. Pei, J., Jiang, D., & Zhang, A. (2005). On mining cross-graph quasi-cliques. In *SIGKDD* (pp. 228–238). ACM.

139. Phillips, C. A., Wang, K., Baker, E. J., Bubier, J. A., Chesler, E. J., & Langston, M. A. (2019). On finding and enumerating maximal and maximum k-partite cliques in k-partite graphs. *Algorithms, 12*(1), 23.

140. Qin, L., Li, R.-H., Chang, L., & Zhang, C. (2015). Locally densest subgraph discovery. In *KDD* (pp. 965–974). ACM.

141. Saito, K., Yamada, T., & Kazama, K. (2008). Extracting communities from complex networks by the k-dense method. *IEICE Transactions on Fundamentals of Electronics, Communications and Computer Sciences, 91*(11), 3304–3311.

142. Sarıyüce, A. E., Gedik, B., Jacques-Silva, G., Wu, K.-L., & Çatalyürek, Ü. V. (2016). Incremental k-core decomposition: Algorithms and evaluation. *VLDBJ, 25*(3), 425–447.

143. Sarıyüce, A. E., & Pinar, A. (2018). Peeling bipartite networks for dense subgraph discovery. In *WSDM* (pp. 504–512).

144. Schmidt, M. C., Samatova, N. F., Thomas, K., & Park, B.-H. (2009). A scalable, parallel algorithm for maximal clique enumeration. *Journal of Parallel and Distributed Computing, 69*(4), 417–428.

145. Seidman, S. B. (1983). Network structure and minimum degree. *Social Networks, 5*(3), 269–287.

146. Shaham, E., Yu, H., & Li, X.-L. (2016). On finding the maximum edge biclique in a bipartite graph: A subspace clustering approach. In *Proceedings of the 2016 SIAM International Conference on Data Mining* (pp. 315–323). SIAM.

147. Shi, C., Li, Y., Zhang, J., Sun, Y., & Philip, S. Y. (2016). A survey of heterogeneous information network analysis. *TKDE, 29*(1), 17–37.

148. Shi, C., Wang, R., Li, Y., Yu, P. S., & Wu, B. (2014). Ranking-based clustering on general heterogeneous information networks by network projection. In *CIKM* (pp. 699–708). ACM.

149. Shi, J., & Shun, J. (2020). Parallel algorithms for butterfly computations (pp. 16–30).

150. Sim, K., Li, J., Gopalkrishnan, V., & Liu, G. (2009). Mining maximal quasi-bicliques: Novel algorithm and applications in the stock market and protein networks. *Statistical Analysis and Data Mining: The ASA Data Science Journal, 2*(4), 255–273.

151. Sözdinler, M., & Özturan, C. (2018). Finding maximum edge biclique in bipartite networks by integer programming. In *2018 IEEE International Conference on Computational Science and Engineering (CSE)* (pp. 132–137). IEEE.

152. Sun, Y., Aggarwal, C. C., & Han, J. (2012). Relation strength-aware clustering of heterogeneous information networks with incomplete attributes. *PVLDB, 5*(5), 394–405.

153. Sun, Y., Han, J., Yan, X., Yu, P. S., & Wu, T. (2011). Pathsim: Meta path-based top-k similarity search in heterogeneous information networks. *PVLDB, 4*(11), 992–1003.

154. Sun, Y., Han, J., Zhao, P., Yin, Z., Cheng, H., & Wu, T. (2009). Rankclus: Integrating clustering with ranking for heterogeneous information network analysis. In *EDBT* (pp. 565–576). ACM.

155. Sun, Y., Norick, B., Han, J., Yan, X., Yu, P. S., & Yu, X. (2012). Integrating meta-path selection with user-guided object clustering in heterogeneous information networks. In *SIGKDD* (pp. 1348–1356). ACM.

156. Sun, Y., Norick, B., Han, J., Yan, X., Yu, P. S., & Yu, X. (2013). Pathselclus: Integrating meta-path selection with user-guided object clustering in heterogeneous information networks. *ACM TKDD, 7*(3), 1–23.

157. Sun, Y., Yu, Y., & Han, J. (2009). Ranking-based clustering of heterogeneous information networks with star network schema. In *SIGKDD* (pp. 797–806). ACM.

158. Tahoori, M. B. (2006). Application-independent defect tolerance of reconfigurable nanoarchitectures. *ACM Journal on Emerging Technologies in Computing Systems (JETC), 2*(3), 197–218.

159. Tan, J. (2008). Inapproximability of maximum weighted edge biclique and its applications. In *International Conference on Theory and Applications of Models of Computation* (pp. 282–293). Springer.

160. Tatti, N., & Gionis, A. (2015). Density-friendly graph decomposition. In *WWW* (pp. 1089–1099).

161. Tomita, E., Tanaka, A., & Takahashi, H. (2006). The worst-case time complexity for generating all maximal cliques and computational experiments. *Theoretical Computer Science, 363*(1), 28–42.

162. Tsourakakis, C. (2015). The k-clique densest subgraph problem. In *WWW* (pp. 1122–1132).

163. Verma, A., & Butenko, S. (2013). Network clustering via clique relaxations: A community based. *Graph Partitioning and Graph Clustering, 588*, 129.

164. Wang, J., & Cheng, J. (2012). Truss decomposition in massive networks. *PVLDB, 5*(9), 812–823.

165. Wang, J., de Vries, A. P., & Reinders, M. J. T. (2006). Unifying user-based and item-based collaborative filtering approaches by similarity fusion. In *SIGIR 2006: Proceedings of the 29th Annual International ACM SIGIR Conference on Research and Development in Information Retrieval, Seattle, Washington, USA, August 6–11, 2006* (pp. 501–508).

166. Wang, K., Cao, X., Lin, X., Zhang, W., & Qin, L. (2018). Efficient computing of radius-bounded k-cores. In *ICDE* (pp. 233–244). IEEE.

167. Wang, K., Lin, X., Qin, L., Zhang, W., & Ying, Z. (2021). Towards efficient solutions of bitruss decomposition for large-scale bipartite graphs. *The VLDB Journal* 1–24.

168. Wang, K., Lin, X., Qin, L., Zhang, W., & Zhang, Y. (2019). Vertex priority based butterfly counting for large-scale bipartite networks. *PVLDB, 12*(10), 1139–1152.

169. Wang, K., Lin, X., Qin, L., Zhang, W., & Zhang, Y. (2020). Efficient bitruss decomposition for large-scale bipartite graphs. In *ICDE* (pp. 661–672). IEEE.

170. Wang, K., Zhang, W., Lin, X., Zhang, Y., Qin, L., & Zhang, Y. (2021). Efficient and effective community search on large-scale bipartite graphs. *ICDE*.

171. Wang, K., Zhang, W., Zhang, Y., Qin, L., & Zhang, Y. (2021). Discovering significant communities on bipartite graphs: An index-based approach. *TKDE*.

172. Wang, R. W., & Fred, Y. Y. (2019). Simplifying weighted heterogeneous networks by extracting h-structure via s-degree. *Scientific Reports, 9*(1), 1–8.

173. Wang, Y., Cai, S., & Yin, M. (2018). New heuristic approaches for maximum balanced biclique problem. *Information Sciences, 432*, 362–375.

174. Wang, Z., Yuan, Y., Zhou, X., & Qin, H. (2020). Effective and efficient community search in directed graphs across heterogeneous social networks. In *Australasian Database Conference (ADC)* (pp. 161–172).

175. Wen, D., Qin, L., Zhang, Y., Lin, X., & Yu, J. X. (2016). I/O efficient core graph decomposition at web scale. In *ICDE* (pp. 133–144). IEEE.

176. Weng, T., Zhou, X., Li, K., Peng, P., & Li, K. (2022). Efficient distributed approaches to core maintenance on large dynamic graphs. *IEEE Trans. Parallel Distributed System, 33*(1), 129–143.

177. Yan, C., Burleigh, J. G., & Eulenstein, O. (2005). Identifying optimal incomplete phylogenetic data sets from sequence databases. *Molecular Phylogenetics and Evolution, 35*(3), 528–535.

178. Yan, X., Zhou, X. J., & Han, J. (2005). Mining closed relational graphs with connectivity constraints. In *SIGKDD* (pp. 324–333). ACM.

179. Yang, J., Peng, Y., & Zhang, W. (2022). (p,q)-biclique counting and enumeration for large sparse bipartite graphs. *PVLDB, 15*(2), 141–153.
180. Yang, Y., Fang, Y., Lin, X., & Zhang, W. (2020). Effective and efficient truss computation over large heterogeneous information networks. In *ICDE* (pp. 901–912). IEEE.
181. Yang, Y., Fang, Y., Orlowska, M., Zhang, W., & Lin, X. (2021). Efficient bi-triangle counting for large bipartite networks. *PVLDB, 14*(6), 984–996.
182. Yu, K., Long, C., Deepak, P., & Chakraborty, T. (2021). On efficient large maximal biplex discovery. *TKDE*.
183. Zeng, Z., Wang, J., Zhou, L., & Karypis, G. (2006). Coherent closed quasi-clique discovery from large dense graph databases. In *SIGKDD* (pp. 797–802). ACM.
184. Zeng, Z., Wang, J., Zhou, L., & Karypis, G. (2007). Out-of-core coherent closed quasi-clique mining from large dense graph databases. *TODS, 32*(2), 13–es.
185. Zhang, F., Xie, J., Wang, K., Yang, S., & Jiang, Y. (2021). Discovering key users for defending network structural stability. *World Wide Web* 1–23.
186. Zhang, F., Zhang, Y., Qin, L., Zhang, W., & Lin, X. (2017). When engagement meets similarity: Efficient (k, r)-core computation on social networks. *PVLDB, 10*(10), 998–1009.
187. Zhang, W., Wang, K., Zhang, Y., & Lin, X. (2020). Cohesive structure based bipartite graph analysis: From motif level to subgraph level. In *DASFAA* (pp. 1–8). Springer.
188. Zhang, Y., & Parthasarathy, S. (2012). Extracting analyzing and visualizing triangle k-core motifs within networks. In *ICDE* (pp. 1049–1060). IEEE.
189. Zhang, Y., Phillips, C. A., Rogers, G. L., Baker, E. J., Chesler, E. J., & Langston, M. A. (2014). On finding bicliques in bipartite graphs: A novel algorithm and its application to the integration of diverse biological data types. *BMC bioinformatics, 15*(1), 110.
190. Zhang, Y., Qin, L., Zhang, F., & Zhang, W. (2019). Hierarchical decomposition of big graphs. In *ICDE* (pp. 2064–2067). IEEE.
191. Zhang, Y., Wang, K., Zhang, W., Lin, X., & Zhang, Y. (2021). Pareto-optimal community search on large bipartite graphs. In *CIKM* (pp. 2647–2656).
192. Zhang, Y., & Yu, J. X. (2019). Unboundedness and efficiency of truss maintenance in evolving graphs. In *SIGMOD* (pp. 1024–1041). ACM.
193. Zhang, Y., Yu, J. X., Zhang, Y., & Qin, L. (2017). A fast order-based approach for core maintenance. In *ICDE* (pp. 337–348). IEEE.
194. Zhang, Z., Huang, X., Xu, J., Choi, B., & Shang, Z. (2019). Keyword-centric community search. In *ICDE* (pp. 422–433). IEEE.
195. Zhou, A., Wang, Y., & Chen, L. (2020). Finding large diverse communities on networks: The edge maximum k*-partite clique. *PVLDB, 13*(11), 2576–2589.
196. Zhou, R., Liu, C., Yu, J. X., Liang, W., Chen, B., & Li, J. (2012). Finding maximal k-edge-connected subgraphs from a large graph. In *EDBT* (pp. 480–491).
197. Zhou, R., Liu, C., Yu, J. X., Liang, W., & Zhang, Y. (2014). Efficient truss maintenance in evolving networks. arXiv preprint arXiv:1402.2807.
198. Zhou, Y., & Hao, J.-K. (2019). Tabu search with graph reduction for finding maximum balanced bicliques in bipartite graphs. *Engineering Applications of Artificial Intelligence, 77*, 86–97.
199. Zhou, Y., & Liu, L. (2013). Social influence based clustering of heterogeneous information networks. In *KDD* (pp. 338–346). ACM.
200. Zhou, Y., Rossi, A., & Hao, J.-K. (2018). Towards effective exact methods for the maximum balanced biclique problem in bipartite graphs. *European Journal of Operational Research, 269*(3), 834–843.
201. Zhou, Z., & Amini, A. A. (2019). Analysis of spectral clustering algorithms for community detection: The general bipartite setting. *Journal of Machine Learning Research, 20*, 47–1.
202. Zhou, Z., & Amini, A. A. (2020). Optimal bipartite network clustering. *Journal of Machine Learning Research, 21*(40), 1–68.
203. Zhu, R., Zou, Z., & Li, J. (2018). Diversified coherent core search on multi-layer graphs. In *ICDE* (pp. 701–712). IEEE.

204. Zhu, R., Zou, Z., & Li, J. (2019). Fast diversified coherent core search on multi-layer graphs. *The VLDB Journal, 28*(4), 597–622.
205. Zhu, Y., He, J., Ye, J., Qin, L., Huang, X., & Yu, J. X. (2020). When structure meets keywords: Cohesive attributed community search. In *CIKM* (pp. 1913–1922).
206. Zou, Z. (2016). Bitruss decomposition of bipartite graphs. In *DASFAA* (pp. 218–233). Springer.

Printed in the United States
by Baker & Taylor Publisher Services